SPANNING TIME
Vermont's Covered Bridges

SPANNING TIME
Vermont's Covered Bridges

To Fred

Joseph C. Nelson

Joseph C. Nelson
12-25-98

THE NEW ENGLAND PRESS, INC.
Shelburne, Vermont

© 1997 by Joseph C. Nelson
Photographs © by Joseph C. Nelson
Illustrations © by Joseph C. Nelson
Designed by Andrea Gray

All rights reserved. No part of this book may be reproduced or transmitted in any form or by any means, electronic or mechanical, including photocopying, recording, or by any information storage and retrieval system, without permission in writing from the publisher, except by a reviewer, who may quote brief passages in a review.

Printed by Regent Publishing Services, Ltd., Hong Kong
First edition

For additional copies of this book or for a catalog of our
other titles, please write:

The New England Press
P.O. Box 575
Shelburne, VT 05482

Nelson, Joseph C.
    Spanning time : Vermont's covered bridges / Joseph C. Nelson.
        p.   cm.
    Includes bibliographical references and index.
    ISBN 1-881535-25-8
    1. Covered bridges--Vermont.   I. Title.
TG 24.V5N45   1997
624'.2--dc21   97-23083
                                        CIP

CONTENTS

Acknowledgments vii
Preface ix
Introduction xi

1 Bennington County Bridges 3
2 The Otter Creek Basin 13
3 The Wooden Bridges of Charlotte 33
4 The Lamoille River and the North Branch 43
5 The Town of Montgomery 67
6 Route 100 in Northern Vermont 79
7 Crossing the Connecticut 93
8 The Lyndon Bridges 101
9 Route 100 in Central Vermont 113
10 Northfield 127
11 The Northern Tributaries of the White River 135
12 Woodstock 149
13 The Windsor Area 163
14 Rockingham to Grafton 175
15 The Deep South 187

APPENDIX A
A Summary of Vermont's Covered Bridges 198

APPENDIX B
A Covered Bridge Glossary... 230

APPENDIX C
The Bridge Truss 240

APPENDIX D
The Bridge Builders 257

APPENDIX E
A Covered Bridge Reading List 261

Index 265

Acknowledgments

The author would like to thank the many individuals who have contributed to this book:

Jan Lewandoski, Gilbert Newbury, and Paul Ide, wooden bridge professionals who kindly read the manuscript with a critical eye, made many suggestions, spun many yarns, and taught the author how the old wooden trusses work.

The writings of historians Richard Sanders Allen, Herbert Wheaton Congdon, Robert L. Hagerman, Victor Morse, and John Spargo, whose love for their subject brought the covered bridges and those who built them to life for me.

Walter and Marie Tedford, for encouraging me to turn what began as a photo journal into a book.

The attending archivists of the Vermont Historical Society Library in Montpelier, for helping me find my way around their research stacks.

The many town historians and the written histories they have produced for their towns.

The folks at the Vermont Agency of Transportation Construction Division in Montpelier, for giving me working space and access to their files on the covered bridge inspection reports.

Francis Howrigan, Fairfield selectman and former senator from Franklin County, for the guided tour of East Fairfield and the basket of garden-ripe tomatoes.

Caroline Brown, president of the Westford Historical Society, for the story of the rescue of the community's bridge.

Susan Girardin and Kerry Young, fifth grade teachers at Camels Hump Middle School, for inviting me into their classroom to see how covered bridges can be used to teach the three Rs.

Sharon Levin, formerly of the Waters Memorial Library, for finding me so many obscure books via the inter-library network.

The Fairfax Library, for giving me access to the treasures collected by the Fairfax Historical Society.

Bea Phillips of North Troy, for leading me to the story of the naming of her town.

And last but hardly least, New England Press editor Mark Wanner, for his guidance, patience, and his unerring "blue pencil."

Joe Nelson
April 2, 1997

Preface

Over the past few decades, Vermont has become widely known for the high quality of life enjoyed by both its residents and its visitors. This Vermont "mystique," as it is sometimes called, is based on its natural beauty and rural character, of course, but there is also a growing appreciation for the way the state has preserved its heritage. Ironically, the preservation has been aided by well-known Yankee frugality, and Vermonters still usually "use it up, wear it out, make it do, or do without," as a local saying goes. Largely because of the enduring desire to "make it do," one can still find quaint villages—complete with white churches, old-fashioned town halls, wonderfully ramshackle general stores, and village greens—that are a world apart from suburban sprawl.

This idyllic picture is not complete, however, unless an old covered bridge happens to be nearby, built of timbers from long-gone forests and serving as a monument to early times. Covered bridges are magnetic to visitors from near and far, and their allure seems to go beyond their obvious charm and historical interest. Perhaps the extraordinary appeal of the bridges rests in their ability to evoke the spirit of the past, transporting viewers back to the days of horses and buggies, sleigh rides, and a simpler way of life.

While viewing a covered bridge is satisfying in itself, most people find themselves wondering about many things afterwards. When and how was the bridge originally built? Why was it covered? How many other bridges are in the area, and what are they like? This book answers these questions and many more. The purpose of *Spanning Time: Vermont's Covered*

Bridges is to celebrate the bridges, show what makes each of them unique, offer some bridge lore and history, and, hopefully, stimulate the reader to want to know more.

To really appreciate the bridges, the reader should know something about them and what to look for. To that end, all of Vermont's covered bridges are organized into fifteen tours in *Spanning Time*, and observations and a brief history are provided for each bridge. In addition, the appendices include descriptions of the several bridge construction types, a glossary of terms, and a master bridge listing. The listing contains a summary of the vital statistics of each bridge, and it also notes superlatives—the highest, the lowest, the longest, the oldest, and so on.

An interesting aspect of exploring the rural areas of Vermont is that, at the time of this writing, many roads are not marked. Because of this, maps and written instructions are included with each tour to provide enough direction to locate even the most remote sites. Still, it would be wise to carry a regular map or road atlas when setting out to view bridges, especially in rural areas. The Vermont Official State Map and Touring Guide is available free from the Vermont Travel Division, Montpelier, VT 05602. *The Vermont Atlas and Gazetteer* by DeLorme and *The Vermont Road Atlas and Guide* by Northern Cartographic are both excellent resources and are both widely available at retail outlets for a modest cost.

For history buffs, a reading list is included. It is by no means a complete list of the many volumes written, but it is a start—most of the books listed also include bibliography or source lists for further reading. Regrettably, some of the books are out of print and may be hard to find. For these, local libraries and historical societies have proven to be very helpful.

Introduction

A few years ago historians were lamenting the passing of the covered bridge and anticipating the time when they would be found only in parks. Fortunately for those who find value in the past, there has been an awakening. Concerned people and local governments are more aware of the value of the old bridges. Preservation is preferred over demolition.

One reason for preservation is tourism. Nostalgia is a great marketing force—come to Vermont for a view of the past, and while you are here, stay at our inns, dine in our restaurants, and buy our maple syrup. A more profound reason for preservation is to maintain links to our past. We can read about our beginnings to learn who we are and where we came from, but the actual places and things provide a sense of scale to make history real.

There is also a purely practical reason for keeping the old bridges—replacing them carries a high price. Thrift is something Vermonters still understand very well. Most of the bridges are located in sparsely settled areas, and, with small tax bases, the townspeople prefer not to pay to build a new bridge. In communities dedicated to agriculture and light industry, the traffic may not require a modern bridge at every crossing. That most of the old bridges have but a single lane is no big problem—villagers would rather wait for an oncoming vehicle to pass than have their taxes increased. The old Yankee saying "If it ain't broke, don't fix it" has currency with bridges.

Of course, as communities grow, traffic increases and industry comes. The Vermont Agency of Transportation finances the repair or restoration of the bridges that are part of the transportation system, but many of the old spans cannot accommodate the demands of modern traffic. Happily, these bridges are no longer knocked down as barriers to progress, and people are increasingly willing to spend the time and money needed to preserve them. Local historical societies work hard to preserve "retired" bridges, and they raise money through community activities, donations, and grants to insure that future generations will inherit them.

Fueling interest in maintaining functional wood bridges is the fact that modern concrete and steel bridges no longer have the glowing reputation they once enjoyed. When the flood of 1927 destroyed nearly two hundred wooden bridges, the state helped the communities replace them with modern multi-lane spans. The new design was thought to be the answer to a prayer for a maintenance-free, once-in-a-hundred-years solution to the problem of crossing rivers and streams. It was not to be so. The new bridges quickly began to succumb to road salt and moisture, some of them after serving but a third of the expected life span. On the other hand, many old wooden bridges continue to serve beyond the century mark, and the Pulp Mill Bridge in Middlebury has been in service since 1820. The secret to long life for a bridge is in keeping the structure dry. Salt does not affect wood, but moisture does—it supports fungus, burrowing insects, and other decay organisms. And that is why the covered bridges are covered—to keep the wooden trusses dry.

A Word About Wood

If wood is good, why do bridge builders almost always use other materials? One reason is that the builders are not as familiar with timber as a material for structural design as they are with steel and concrete. The cement and steel industries have long done their homework and the timber industry has not. The typical engineer is familiar with the compression, tension, and shear characteristics of concrete and steel because those industries have long been promoting their products by teaching potential customers how to use them.

Wood, having demonstrated its potential for long life in bridge construction, has a role to play in modern bridge design, what with the aging infrastructure (read bridges) of the nation falling apart. The Forest Service, under the U.S. Department of Agriculture, is working with the timber industry in a cooperative program to promote timber technology, educate engineers, develop new bridge designs, and to share new timber technology with users.

If new wooden bridges take their places over the nation's streams, they will not all be covered bridges like in old times. The old timbers will be supplanted with weather-proof pressure-laminated wooden beams, decks, and arches using new glues and preservatives. An example of the new "glulam" beam technology can be seen in the recently renovated Kidder Hill Bridge in Grafton.

How Many Covered Bridges Are There?

Bridge historian Richard Sanders Allen wrote in *Covered Bridges of the Northeast*: "If it had not been for the '27 flood, Vermont would have many more covered bridges than are standing today. The exact number has never been compiled, but it is estimated that nearly two hundred covered wooden spans in the Green Mountain State were destroyed."

There were more than five hundred wooden bridges before the big flood. The number of covered bridges surviving today depends upon who is doing the counting. Counts differ with the definition of what a covered bridge is. A liberal definition is anything that spans a stream and has a roof over it is a covered bridge. A conservative definition requires the bridge trusses to be complete and functional, unaided by reinforcement with modern materials and design, and the bridge should have an historical context—that is, it must be old. The counts vary from bridge historian Neal G. Templeton's 114 to seventy-five, the Vermont Agency of Transportation's count, which only includes the public covered bridges that are in service.

Spanning Time includes all of the historic bridges, whether reinforced or not, plus the replicas constructed using authentic materials and design. Using this definition, in 1997 there were 101 covered bridges in the state. Of these, three are owned and maintained by the state of New Hampshire, eight are privately owned, and three are replicas. Twenty-six of the bridges in use have been strengthened for modern traffic, leaving the truss redundant engineering-wise but worth viewing for history's sake. The rest, no longer carrying traffic, are bypassed or have been moved to parks. (See the summary chart at the end of Appendix A.)

The Future of the Covered Bridges in Vermont

The number of bridges in use will likely change. The Vermont Agency of Transportation has recently completed an inspection of seventy-five of the covered bridges that serve state and town highways. In the course of the survey, some of the bridges were found to be unsafe and were closed to traffic—others were found to be in need of repair. The towns, ultimately responsible for a share of the funding, will decide what work will be done.

The inspection program was part of a long-range plan to oversee public safety, plan for current and future traffic needs, and preserve all covered bridges in the state. The structure of each bridge was inspected for safety, and the bridge traffic was evaluated. The results of the study were turned over to the local communities with recommendations to help them decide whether to repair, rehabilitate, or replace their covered bridges.

The recommendations include one of five options based on the condition of the bridge and the type of traffic the bridge supports. The community may close the bridge to vehicular traffic, with traffic diverted to existing roads and bridges; continue to use the bridge for light vehicular traffic, with heavier truck traffic diverted to other routes; close the bridge to traffic and construct an adjacent bypass bridge; rehabilitate the bridge to support moderate traffic safely, or replace the bridge and move it to a nearby preservation site.

The inspections were completed in the spring of 1995. Most of the bridges have passed the structural inspection and continue to be in use. Hopefully, they will be with us for a long time to come. Now, go see them.

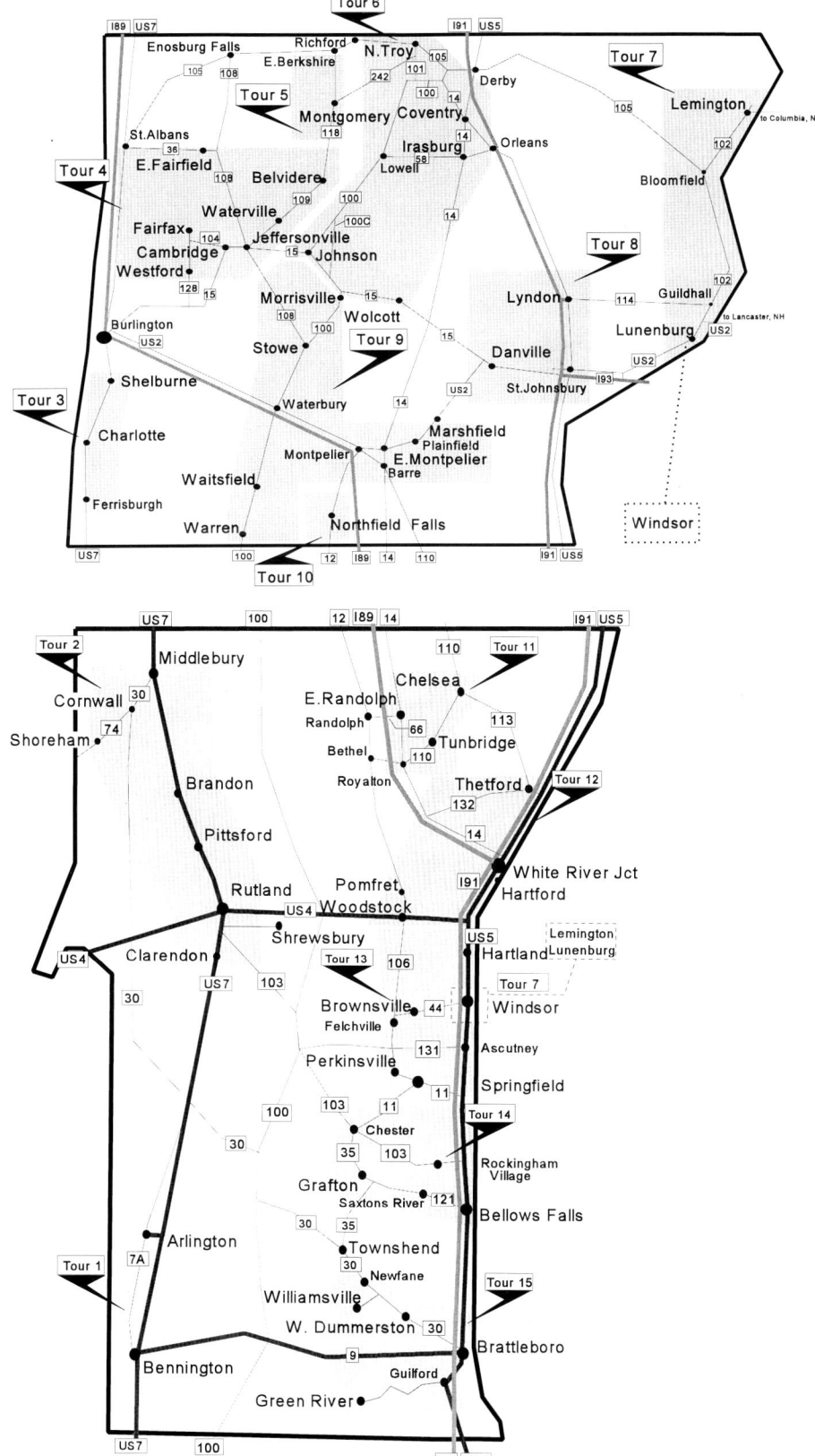

BENNINGTON COUNTY BRIDGES

1

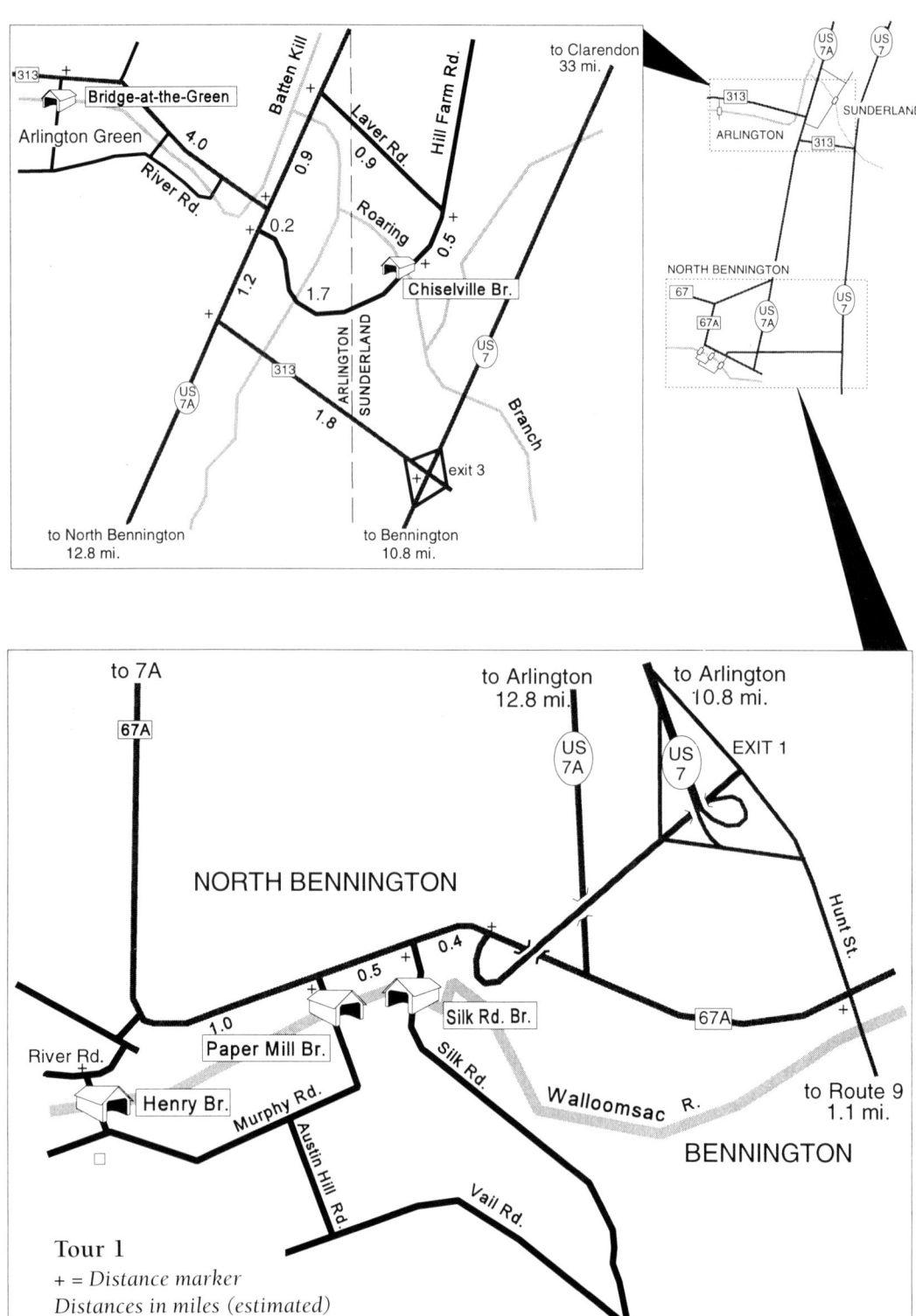

BENNINGTON COUNTY BRIDGES

Tour 1

Bennington County is home to five of the most picturesque covered bridges to be found anywhere. To the west of Arlington Village stands the celebrated Bridge-at-the-Green—to the east in Sunderland is the Chiselville Bridge. Further south in North Bennington are the Henry Bridge, the Silk Road Bridge, and the Paper Mill Bridge.

All five bridges are easily reached from U.S. Route 7, the scenic route connecting the towns of western Vermont. The traveler entering the state from south or north by way of Route 7 will pass close to many of the village greens and old white meeting houses made famous by calendar and post card.

BENNINGTON

Named for Benning Wentworth, Royal Governor of New Hampshire, Bennington was chartered as a New Hampshire town in 1749 and settled in 1761. Blessed with rivers and streams fed by the Green Mountain watershed, Bennington flourished and early industry prospered. Beginning with gristmills and sawmills, the valley villages soon became home to paper and textile factories. With the growth of the communities came the ubiquitous covered bridges to connect them. Many wooden bridges were built before the coming of steel and concrete, but of those built, only five remain.

Postcards and calendars have made the Bridge-at-the-Green perhaps the best known covered bridge in Vermont. There is a great swimming hole here, and the base of the abutment provides a popular jumping-in place. Notice the cables, installed for lateral bracing.

Bridge-at-the-Green

Arlington Green is one of those picture-perfect New England town greens that artists and photographers seek out. Norman Rockwell, one of America's most popular painters, once lived and worked nearby. The green is found off Route 313, four miles west of the junction with U.S. 7 in Arlington Village.

Arlington Green lies across the famed Batten Kill trout stream from Route 313, reached by an eighty-foot plank-lattice bridge that has been serving travelers since 1852. The Chapel-on-the-Green stands beyond the bridge's south portal, its square tower rising above the trees.

The Bridge-at-the-Green demonstrated the sturdiness of Ithiel Town's patented plank-lattice truss. Swept off its abutments by a flood on the Batten Kill in the early 1850s, it came to rest on its side intact and still spanning the stream. Because it was the only bridge nearby that

> **Bridge-at-the-Green**
>
> TOWN: Arlington
> DATE: 1852
> TRUSS LENGTH: 80'4"
> BUILDER: Unknown
> STRUCTURE: Plank-lattice

still crossed the river, people continued to use it, crossing over on the upturned side.

Folklore has it that after the bridge was disassembled and returned to its proper attitude, it was lashed into place with cables to keep it from going astray again. It is true—the cables are there to give lateral support to the structure against wind and water. Except for the cables, the bridge remains structurally original.

Chiselville Bridge

The Chiselville Bridge, also much photographed and painted, lies one mile east of Arlington Village, in Sunderland. The easiest way to find it is to take Laver Road where it leaves Route 7 east, 0.9 miles north of the junction with Route 313. At Hill Farm Road, turn south and proceed 0.5 miles. An alternate way is to take the first eastbound road leaving Route 7 just south of the junction with Route 313.

The Roaring Branch of the Batten Kill lived up to its name when it destroyed the bridge spanning the mill pond above Chiselville in 1869. The flood inundated the

Standing forty feet above the stream bed, the Chiselville Bridge is well out of reach of the Roaring Branch. This span has been known to inspire vertigo.

BENNINGTON COUNTY BRIDGES

Sunderland flats and took a heavy toll on area roads and bridges.

Because the Roaring Branch and its forks carry much of the runoff from the west face of the Sunderland portion of the Green Mountains, it was thought that a new bridge built at the original site would likely meet a similar fate. The following year Daniel Oatman built a new plank-lattice bridge on the crags forty feet above the stream, spanning 117 feet. Chiselville Bridge is the second highest covered bridge above a stream bed in Vermont.

The Agency of Transportation strengthened the bridge in 1973. After 103 years of service, the bridge's age and the modern traffic it handled made the original construction unsafe. The structure now consists of an independent slab roadway supported by three steel beams and a concrete pier. The trusses put in place by Oatman now support only themselves and the winter snow-load.

The Chiselville Bridge has cables similar to those at the Bridge-at-the-Green to brace it against the winds that howl through the gorge.

> **Chiselville Bridge**
> TOWN: Sunderland
> DATE: 1870
> LENGTH: 116'6"
> BUILDER: Daniel Oatman
> STRUCTURE: Plank-lattice

NORTH BENNINGTON

North Bennington's three covered bridges cross the Walloomsac River where it flows parallel to Route 67A. This stretch of highway connecting North Bennington and Bennington can be extremely congested, so it is wise to avoid visiting the bridges at certain times of day. Once you cross one of the three bridges, all of them can be reached by less busy back roads.

The Henry Bridge, westernmost of the three, is found on River Road. River Road leaves the south side of Route 67A, where that highway bends sharply east. Silk Road serves the easternmost bridge, while the Paper Mill Bridge in the middle is accessed by Murphy Road.

Henry Bridge

The appearance of the Henry Bridge is striking with its red paint and white trim. Open ports along both sides of its 125-foot length expose the plank-lattice truss, giving the bridge a natty look. A signboard in the gables

> ### The Standoff at the Bridge
>
> In July 1771 word was received that a New York sheriff and his posse were on their way to take possession of disputed property in the Bennington area. The settlers prepared a reception–about six men awaited the posse at the bridge over the Walloomsac near William Henry's homestead, while many more men hid in the woods and fields around the bridge. When the advance party of the posse reached the bridge, they were allowed to meet with Mr. Breckenridge, a Green Mountain Boy leader, to plead their case. After the main body of the posse arrived, the settlers at the bridge informed the sheriff that they would not leave their land voluntarily. When the sheriff ordered his posse to advance, the hidden men stepped forward with muskets ready. The sheriff quickly concluded that discretion was the better part of valor, and the posse retreated back over the bridge. The bridge was also used in 1777 by troops on their way to the Battle of Bennington.

> ### Henry Bridge
>
> TOWN: Bennington
> DATE: 1989
> LENGTH: 120'6"
> BUILDER: VAOT
> STRUCTURE: Plank-lattice

reads "c1840," but the bridge you see today is a replica built in 1989 by the Vermont Agency of Transportation.

The bridge, named for a family long in residence nearby, crosses the Walloomsac River below a dam and a large mill pond. The spot is a historic one—the crossing here has been important since early colonial times, and it served as the setting for one of the Green Mountain Boys' bloodless "victories" over New York authorities.

The first covered Henry Bridge became known as the strongest in New England when the Burden Iron Company of Troy, New York, began operations in the 1860s on a neighboring piece of land containing iron ore. The company strengthened the span by adding another set of plank-lattice trusses to each side to support ore wagons.

Ironically, the additional trusses did not strengthen the bridge, but merely added to the weight of the span and created crevasses that collected moisture-absorbing dirt. The iron company's structural engineering expertise apparently stopped at the "more must be better" school of thinking. The Town of Bennington removed the extra trusses from the old Henry when restoration began on its three covered bridges in 1952. The replica has a single set of trusses as the original builder intended.

The Vermont Agency of Transportation posted a sign by the replica bridge that nutshells some history of the area:

HENRY COVERED BRIDGE
ACROSS THE WALLOOMSAC RIVER

This quiet spot was once a major river crossing. Traffic between southwestern Vermont and New York State crossed here until a railroad was built in 1852. Troops marched from Manchester, Vermont to the Battle of Bennington in 1777 and teams and stages transported freight and passengers. The original Henry Bridge was built c. 1840. In the late 1860's and '70s heavy wagon loads of iron ore were hauled over the bridge from the Burden Iron Company mine on Ore Bed Road to its washing works on Paran Creek in North Bennington. A succession of water-powered mills was located next to the bridge on the south side.

The last was a gristmill operated into the 1920s by Berntine T. Henry, one of this area's many descendants of the Irish born William Henry (1734-1811). . . . This bridge, built in 1989 by the State of Vermont Agency of Transportation, is a replica replacing the deteriorating original bridge built c. 1840.

Two other covered bridges, the Paper Mill Village Covered Bridge and the Silk Road Covered Bridge, cross the Walloomsac River within two miles upstream.

The new Henry Bridge is a monument to all of the bridges that have spanned the Walloomsac River at this location. The crossing here has been important since early colonial times.

Swimmers have long enjoyed plunging off the roof of the Paper Mill Bridge into the mill pond below. In the middle 1990s the roadway was blocked off to pedestrians pending reconstruction.

Paper Mill Bridge

The Paper Mill Bridge spans a mill pond and often makes a striking reflection on the clear, still water. The low rambling mill buildings lining the shore share the reflection. The bridge is handsome in red paint and white trim, and the open ports along both sides of its 125-foot length expose the plank-lattice truss and bathe the interior with a quiet warm light.

Built in 1889, the old bridge was eventually bypassed by a temporary single-lane iron span. It continued to serve strollers and cyclists until 1995, when it was blocked against all traffic by the town. A Vermont Agency of Transportation inspection had found it to be critically deteriorating and on the verge of collapse. The inspection team recommended that the bridge be rehabilitated for moderate traffic.

The casual viewer might notice the Paper Mill Bridge bears a family resemblance to the bridge on Silk Road. That both were restored in 1952 and are maintained by the Town of Bennington explains part of this. Under the paint and trim, however, the fundamental similarity of the structures may be derived from the possibility that

---

Paper Mill Bridge

TOWN: Bennington
DATE: 1889
LENGTH: 125'2"
BUILDER: Charles F. Sears
STRUCTURE: Plank-lattice

they were built by father and son. According to the Vermont Division of Historic Sites, the Paper Mill Bridge was built by Charles F. Sears, and the Silk Road Bridge was probably built by Benjamin Sears, father of Charles (see below).

Silk Road Bridge

The Silk Road Bridge crosses the Walloomsac River in a suburban setting. Well maintained and in daily use, it is believed to have been built by Benjamin Sears, circa 1840. It was restored in 1991 by Gilbert Newbury. Open ports along most of its ninety-foot length expose the web of the plank-lattice truss. On bright days, the interior is traced with geometric shadow patterns, changing with the hour and with the season.

Mr. Newbury relates that during the restoration, the crew befriended a young boy who often came to watch the work with his grandfather. The crew foreman etched the boy's initials and date into the top of a treenail and helped the boy drive it in place. The treenail can be found somewhere in the western truss.

Silk Road Bridge
TOWN: Bennington
DATE: c. 1840
TRUSS LENGTH: 88'3"
BUILDER: Benjamin Sears
TRUSS TYPE: Plank-lattice

There has been a bridge on the site of the present-day Silk Road Bridge since the 1790s, if not earlier. Over the years the bridge has also been called the Locust Grove Bridge and the Robinson Bridge.

> ### What's in a Name?
>
> Was the name of Bennington's river first given by the Algonquian tribes who once inhabited the region? Not according to two stories found in the library of the Vermont Historic Society.
>
> "At first the name of the stream was spelled Walloonsac," wrote Alexander B. R. Drysdale in his *Bennington's Book* (1927). It was so called, Drysdale wrote, in honor of an eccentric Dutchman named Van Vetchen Van Der Spiegel, who lived on the banks of the river below Bennington. Van Der Spiegel raised rabbits known as Walloon hares. He used to sell them door to door to the settlers, carrying them about in a sack. Supposedly, he became known as Walloon-sack Van Der Spiegel, and the stream he lived near was called Walloonsack's River.
>
> The other name-source comes from John and Caroline Merrill in their *Sketches of Historic Bennington* (1898). According to the Merrills, Walloomsac comes from the Dutch word Wallumschaik, "chaik" meaning scrip, or patent. The name Walloomsac, then, comes from "Wallum's Patent," the name of a grant issued by New York on June 15, 1739, about ten years before the charter under New Hampshire.
>
> The second explanation seems reasonable, but the first is more entertaining. It is appealing to think that an eccentric old man peddling his hares in a sack could leave his mark on area place-names.

The span has also been known both as the Locust Grove Bridge and the Robinson Bridge. There was another bridge named Robinson that crossed the Walloomsac in sight of the old Governor Robinson mansion. That bridge was reputed to have been haunted, but it and its resident spirits disappeared long ago.

Sources

Allen, Richard Sanders. *Covered Bridges of the Northeast.* Brattleboro, VT: Stephen Greene Press, 1957.

Hemenway, Abby Maria. *The Vermont Historical Gazetteer, Volume I.* Burlington, VT: A. M. Hemenway, 1868.

Spargo, John. *Covered Wooden Bridges of Bennington and Vicinity.* Bennington, VT: Bennington Historical Museum and Art Gallery, 1953.

Vermont Agency of Transportation Covered Bridge Study. Prepared for the State of Vermont Agency of Transportation by McFarland-Johnson, Inc., Binghamton, New York, 1995.

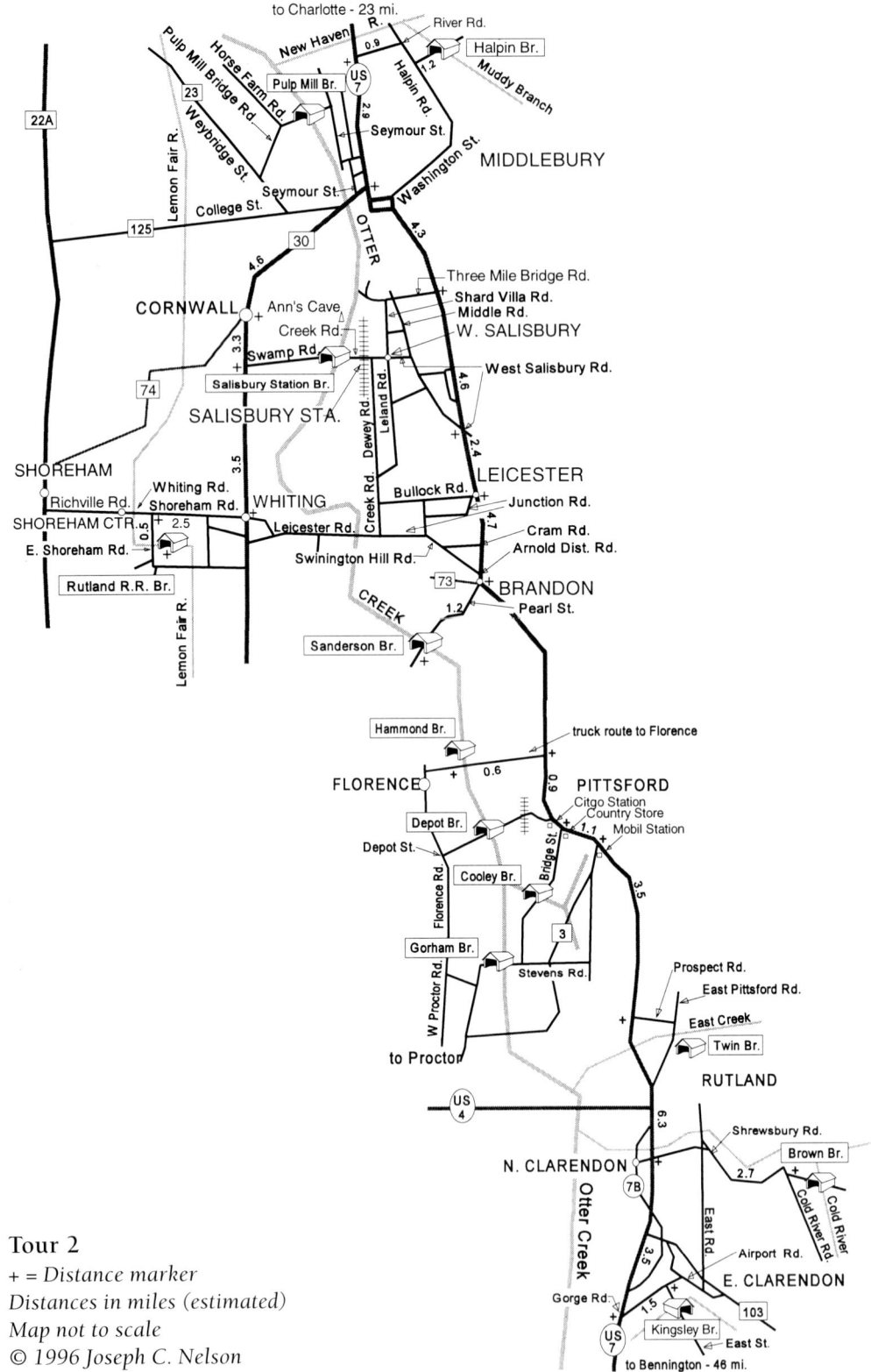

Tour 2
+ = *Distance marker*
Distances in miles (estimated)
Map not to scale
© *1996 Joseph C. Nelson*

The OTTER CREEK BASIN

Tour 2

Otter Creek flows north about seventy miles from Dorset, in the Green Mountains, to Lake Champlain in Ferrisburgh. Eleven covered bridges are found along a fifty-mile stretch of Otter Creek, the longest river within Vermont's borders. Much of the way is accessible from scenic Route 7 and the paralleling Route 30. Adventurous travelers can take side-trips along some of the Otter's many tributaries.

The Middlebury Area

John Everts contracted to survey three towns in the 1760s. He named the one to the south Salisbury and the one to the north New Haven. He named the third town Middlebury for its position between the two. The town was incorporated in 1816 as "Middlebury Borough." The name was changed in 1852 to the Village of Middlebury.

In the covered-bridge world, Middlebury is known for its Pulp Mill Bridge, one of the last two-lane arch bridges in Vermont. Built for the Waltham Turnpike Company in or around 1820[1], it is possibly the oldest Vermont span still standing. There are two other bridges close by, less well known but worth visiting. To the north, the Halpin Bridge sits beside the Middlebury-New Haven town line. To the south, the Station Bridge connects Salisbury to Cornwall.

[1] This date is disputed. Details are provided later in the text.

Halpin Bridge

The sixty-six-foot Halpin Bridge crosses forty-one feet above the New Haven River's Muddy Branch, the highest any Vermont covered bridge rests above a stream bed. Built in 1824 by unknown craftsmen, it is also one of the state's oldest surviving plank-lattice bridges still in use.

In 1994, after state bridge inspectors found the abutments to be dangerously cracked, the plank-lattice structure was lifted from its crossing place by crane so repairs could be made. The bridge itself was found to be in good structural condition, so, after replacement of some of the bottom chord planks and lattice members, it was moved back over the stream, resplendent with new siding and roof.

The original abutments, now replaced with cast concrete, were remarkable in that they were not only very tall, they were built from cut white marble blocks. Similar masonry can still be seen to the north of the bridge in some abandoned foundations. The bridge was flanked by a mill and a general store in the middle 1800s.

To find the bridge, take River Road where it leaves Route 7 just south of the bridge crossing the New Haven River,

The Halpin Bridge is set forty-one feet above the Muddy Branch of the New Haven River, higher than any other covered bridge in Vermont. The original abutments were marble, laid up dry. The use of marble as a foundation stone indicates that there is a quarry nearby. The town-maintained span leads only to a farmyard.

Halpin Bridge

Town: Middlebury
Date: 1824
Truss length: 66'4"
Builder: Unknown
Truss type: Plank-lattice

about three miles north of the junction of Routes 7 and 30 in Middlebury. Drive east on River Road one mile, then turn right onto Halpin Road. Drive another mile and turn left onto an unmarked dirt road. The town-maintained bridge is in use, but it leads only to a farmyard.

Pulp Mill Bridge

When you visit the Pulp Mill Bridge, take your jogging shoes. The old "double-barrel" bridge stands in a tight little urban neighborhood and parking for bridge viewers is scarce, so visiting the bridge usually involves a bit of a walk.

The Pulp Mill is the only two-lane covered bridge in Vermont that carries regular daily traffic. Joggers and cyclists also compete with the automobiles. Vermont's other two-laner originally served Cambridge. It now stands next to Route 7 in Shelburne and is used as an entry for staff at the Shelburne Museum. This tally does

The Pulp Mill Bridge is one of the oldest covered bridges in Vermont, but its exact date of construction is unknown.[2] Found to be safe by the covered bridge inspection team, the span will continue to serve the Middlebury area for many years to come.

[2]The Agency of Transportation inspection report cites a "purported" date of 1808 but settles for 1820, the date offered by historian H. W. Congdon. Jan Lewandoski, after researching town records, believes the bridge could not have been built prior to 1850.

THE OTTER CREEK BASIN

not count Windsor's two-laner over the Connecticut River, because 99 percent of it is in New Hampshire.

The 184-foot Pulp Mill Bridge once crossed the Otter Creek with a single span. According to the Vermont Agency of Transportation covered bridge inspection report, the two concrete piers topped with timber cribs we see today were built in 1979. When the bridge was subdivided into three spans, the direction of half the braces had to be reversed. The stone abutments were capped and faced with concrete and the laminated arches rebuilt. Additional work was done in 1991 by Jan Lewandoski.

When the Pulp Mill Bridge was constructed, the builders overlooked several key details. The posts, which receive heavy braces, do not shoulder within the chords. That is, the notched posts aren't fitted into lower chord members notched to receive them, latching them in place. Instead, the bottoms of the posts are reduced in thickness to fit between the chord members and held there with a single bolt. The tremendous load of the nearly two-hundred-foot-long span transmitted through the braces set at the bottoms of the posts would break the posts, sliding them between the chord members. The bridge started failing right away, and carpenters have been fixing it ever since.

Early efforts were made to strengthen the truss with Burr-style segmented timber arches—when the arches were added is uncertain. In the 1860s these were augmented with laminated arches of three-inch planks, using the original arch as a form.

Today, there are two large arches between the lanes that rise nearly to the ridge pole. A smaller arch rises to the eaves on each side. The center arches, consisting of a lamination of ten three- by six-inch planks, are bolted to both sides of a central multiple-kingpost truss. The outer arches are a lamination of nine three- by six-inch planks each bolted to the inside of a multiple-kingpost truss. There is no longer any evidence of the original segmented timber arches.

There is access to the creek at the northwest side of the bridge approach. From there, at low water, the entire length of the bridge with piers and abutments can be viewed from the dam. Note that the ends of the arches are bedded in the abutments below the main stringers. Most of the other Vermont bridges built in the "Burr

> **Pulp Mill Bridge**
>
> TOWN: Middlebury-Weybridge
> DATE: 1820
> TRUSS LENGTH: 183'7"
> BUILDER: Unknown
> TRUSS TYPE: Burr arch

manner" terminate the ends of the arches at the bottom chords above the abutments.

Besides being on the National Register of Historic Places, the Middlebury town plan identifies the bridge as a Scenic Roads Resource, giving it additional preservation status. The Town of Weybridge was awarded a grant to fund a pedestrian bridge to be built either beside the covered bridge or attached to it.

The Pulp Mill Bridge was named for a wood-pulp mill that operated nearby. It is located on Weybridge Road on the Weybridge-Middlebury town line and is maintained by both towns. To find the bridge from Route 7, take Route 30 south in Middlebury. Drive about 0.3 miles to Route 23. Turn right onto Route 23 and drive 0.8 miles to Pulp Mill Bridge Road. Turn right to the bridge at the junction of Horse Farm Road.

Salisbury Station Bridge

The 154-foot Salisbury Station Bridge spans Otter Creek on Swamp Road at the edge of the Great Cedar Swamp. It has been known as the Cedar Swamp Bridge or simply as the Station Bridge.

The coming of the railroad to Salisbury provided the incentive to build the Salisbury Station Bridge in this remote spot. The midstream pier was added in 1969 after the span had been in service for 104 years.

> ### Ann Story's Cave
>
> The site of Ann Story's cave lies about one-third of a mile north of the Salisbury Station Bridge. Ann Story's husband Amos had taken rights to one hundred acres not far south of where Swamp Road crosses the railroad tracks today. There, he and his son Solomon built a cabin and cleared a field. Amos was killed by a falling tree, but Ann was determined that the family was going to live on their land. In 1774 she and Solomon and her other four children came to the cabin Amos had built. There the Storys planted the field. In 1775, in the beginnings of the Revolutionary War, Tories and Indians began pillaging unprotected settlements. Other settlers retreated to Rutland, but Ann and her children stayed on to harvest their crops. They dug a tunnel into the west bank of the Otter Creek and stayed there at night while they continued to work on the land by day.
>
> The tunnel remained undiscovered until a Tory heard a baby crying as he passed nearby. He surprised the Storys when they came out and demanded that they leave. Ann sent Solomon to Rutland to warn the people that the Tories were back. As a result, fourteen Tories were captured and taken to Fort Ticonderoga. This, and her resolve to defy the king's forces, gave Ann Story her reputation as a heroine of the revolution. There is no road to the little clearing by Otter Creek, but there is a monument there.

Raised in 1865 by unknown builders, the truss is unique in that the web is wider than that of any other plank-lattice truss in Vermont. It measures four feet ten inches on center—the others average about three feet. In 1969 the bridge was renovated and a mid-stream pier added to support the long span. Until then, the Salisbury Station Bridge was one of the longest single spans in the state. Further work was done in the winter of 1992, when Jan Lewandoski made repairs to the truss.

The casual observer today may wonder why the bridge was built in a remote spot between a large swamp and acres of fields. In the 1860s, however, it provided Cornwall with a vital economic link to the railway shipping point at Salisbury. There is a ford here, indicating that there was a crossing long before there was a bridge.

The best way to reach the Salisbury Station Bridge in most seasons is from Route 30. Look for Swamp Road approximately eight miles south of the junction of Route 7 and Route 30 in Middlebury. Drive east on Swamp Road for two miles—the bridge crosses the town line at Otter Creek. The bridge is co-owned and maintained by the towns of Cornwall and Salisbury.

> ### Salisbury Station Bridge
>
> **Town:** Salisbury-Cornwall
> **Date:** 1865
> **Truss length:** 153'10"
> **Builder:** Unknown
> **Truss type:** Plank-lattice

Shoreham

Shoreham is justly known for its apple orchards. The town is also known, at least among bridge seekers, for the covered railway bridge spanning the Lemon Fair River.

Rutland Railroad Bridge

The Rutland Railroad Bridge is easily missed. It crosses the Lemon Fair River behind a fishing access park and almost out of sight down a tree-lined lane that was once the railroad right-of-way. A historic site marker found at the back of the fishing access parking lot reads:

The Rutland Railroad Bridge once connected Leicester Junction with the Delaware and Hudson Railroad at Ticonderoga, New York. A monument to structural engineering, the old railroad bridge is now a popular place for fishermen.

> **SHOREHAM BRIDGE**
> This 108' Howe Truss railroad bridge is one of only two covered railroad bridges left in Vermont. It was built in 1897 on the 15.6 mile Addison Branch connecting the Rutland Railroad at Leicester Junction with the Delaware and Hudson at Ticonderoga, N.Y. crossing Lake Champlain on a floating bridge at Larrabee's Point. This bridge was last used for railroad traffic in 1951.

> ### The Origin of the Lemon Fair
>
> Lemon Fair is a curious name for a river in Vermont, but the origin of the name has been obscured over the years. Even Abby Hemenway's wonderful *Vermont Historical Gazetteer* provides little help. One local history expert—a surveyor who reads extensively about local history in his land research—has found a plausible origin of the name. He discovered that Lemon Fair may be a corruption of the words "A Lamentable Affair," a community named in commemoration of an Indian attack during the early settlement years.
>
> Many other theories have been proposed, but perhaps the most likely source of the name is that it is an English corruption of a French phrase describing a sometimes murky stream. The river flows over beds of limestone and through soils containing concentrations of hydrate of magnesium sulfate, or epsom salts. There are enough salts in the area to flavor some of the nearby wells.

The bridge, owned by the State Division for Historic Preservation, is on the National Registry of Historic Places as an engineering landmark.

The Rutland Railroad Bridge is best reached from Route 30 at the village of Whiting. Drive west on Shoreham and Whiting roads 2.5 miles, turn south on East Shoreham Road, and proceed 0.5 miles. Look to the left for a state fishing-access park just north of the steel bow-truss bridge crossing Richville Pond.

> ### Rutland Railroad Bridge
>
> TOWN: Shoreham
> DATE: 1897
> TRUSS LENGTH: 109'
> BUILDER: Rutland Railroad
> TRUSS TYPE: Howe

BRANDON

Brandon began in 1761 as Neshobe, named by the Massachusetts proprietors who got their grant from New Hampshire Governor Benning Wentworth. The Vermont legislature changed the name to Brandon in 1784.

Stephen Arnold Douglas, U.S. Senator from Illinois from 1847 to 1861 and Democratic candidate for president against Abraham Lincoln, was born here in 1813 to a prominent family. He left Vermont when he was twenty years old.

Sanderson Bridge

The Sanderson Bridge stands amid cornfields in the fertile bottom lands of the Otter Creek flood plain. It is named for a family that has farmed the adjacent land

> **Sanderson Bridge**
>
> TOWN: Brandon
> DATE: c. 1838
> TRUSS LENGTH: 132'
> BUILDER: Unknown
> TRUSS TYPE: Plank-lattice

Some Brandon historians believe Nicholas Powers may have built Sanderson Bridge. The gable treatment, however, is not typical of Powers' known work. Notice the partial cornice returns and the pilaster moldings.

continuously since 1825. The planking of the ancient bridge has been bleached silver by the coming and going of the seasons. The interior presents a show of geometric shadows and spears of light when the sun shines through the gaps and cracks in the siding, highlighting the plank-lattice truss.

The bridge is remarkable both for its abutments and its gable-ends. The abutments are large white slabs of marble laid dry. The gables are notable for the neo-Grecian treatment of the eaves and gable ends, enclosed cornices, enclosed roof gable-end overhangs, and partial cornice returns. The portal features pilaster moldings. Because of the style of finish used, it is believed the bridge was built in the late 1830s or early 1840s. Among all of Vermont's surviving bridges, only Downers Bridge in Weathersfield, built in the same decade, shares these features.

Designed for nineteenth-century loads, the 132-foot span that once served the Brandon to Sudbury Road was closed with a concrete barrier in the 1980s and bypassed by a bridge of salvaged I-beams. A new steel bridge is slated for construction. The Town of Brandon plans to

THE OTTER CREEK BASIN

rehabilitate the covered bridge for pedestrian and bicycle traffic.

Once known as the Lower Bridge, the Sanderson stands 1.2 miles from Route 7 on Pearl Street on the northern outskirts of Brandon. The Upper Bridge, or Dean Bridge, once stood on Carver Street, crossing Otter Creek upstream of the Sanderson Bridge. Built in the same decade as the Sanderson, it is one of Vermont's lost treasures, destroyed by arsonists.

Pittsford

The Town of Pittsford was granted to Ephraim Doolittle and sixty-three others in 1761. A branch of the old Crown Point military road crossed Otter Creek here, the best fording place in the neighborhood. It was named Pitt's Ford in honor of William Pitt, then Prime Minister of England. Four covered bridges and a bridge site are found in the Town of Pittsford alone, all within a two-mile stretch as the crow flies. The Hammond and Depot bridges are near Florence. Heading south, the abutments of the Mead Bridge, then the Cooley and the Gorham bridges can be viewed. Of the four bridges surviving in the town, three are in use—the Hammond span is retired.

The Pittsford bridges have a common appearance, with a long gable overhang at the portals. The feature is designed to keep moisture away from the bridges' main stringers where they are bedded on the masonry of the abutment. The Gorham, Hammond, Depot, and Cooley bridges all feature this exaggerated portal overhang, and they all use the plank-lattice truss.

No one builder seems to have created the consistent appearance of the bridges. Abraham Owen used it when he built the Depot Bridge in 1840, and again in 1841, when he and Nicholas Powers built the Gorham Bridge. Asa Nourse used it when he built the Hammond Bridge in 1842. Powers employed it in the Cooley Bridge in 1849.

Powers later became a leading covered bridge builder, but the Pittsford overhang is not found elsewhere. Powers' other bridges include Rutland's Twin Bridges, built in 1850 and 1851, and Shrewsbury's Brown Bridge, con-

The Hammond Bridge connected Florence Station with the state highway until it was bypassed with an impressive concrete and steel span. Standing in a grove of trees apart from modern hustle and bustle, the retired bridge inspires dreams of the past.

structed in 1880. None of these bridges show the exaggerated overhangs found in Pittsford.

Hammond Bridge

Standing in the shade of ancient trees, with vines climbing its weather-stained side, the Hammond is a picture-book bridge. The setting inspires one's imagination—the echo of tramping hooves and rolling iron-tired wheels can still be heard if you listen hard enough.

The 139-foot Hammond Bridge spans the Otter Creek. It was lifted off its abutments by the 1927 flood and left stranded in a field more than a mile downstream. The town dragged the bridge back to its proper place during the following winter, when the ground was firm and the ice on the creek would support the work.

The Hammond Bridge is out of service, and the road was rerouted to bypass it. A Historic Site sign stands near

Hammond Bridge

Town: Pittsford
Date: 1842
Truss length: 138'8"
Builder: Asa Nourse
Truss type: Plank-lattice

THE OTTER CREEK BASIN

> ### A Bridge Too Far
>
> Richard Sanders Allen, in his *Covered Bridges of the Northeast*, relates the following anecdote: It was Avery Billings of nearby Rutland who went on record with the classic answer to a petition for one more bridge in his town: "Mr. Moderator!" he bellowed in March Meeting. "In as much as we have already built bridges over Otter Creek at Goodkins Falls, Ripley's, Dorr's and Billing's, four bridges in two and a half miles, Mr. Moderator I move we bridge the whole [dang] creek, lengthwise!"

a portal. To find the bridge, leave Route 7 west on the truck route to Florence 6.7 miles south of Brandon, or 1.1 miles north of the center of Pittsford. Drive 0.6 miles. To see the Depot Bridge, pass the Hammond, crossing the railroad tracks. Drive up the hill to the Florence-Proctor Road. Turn left and continue to a fork in the road. Take the left fork and drive under the narrow old railroad overpass and continue on the Florence-Proctor Road for about 0.9 miles. Turn left on Depot Street to approach the bridge.

Depot Bridge

The Depot Bridge serves Depot Street, east of the village of Pittsford. It crosses Otter Creek in the midst of the bottom lands lying midway between the village and the Florence-Proctor Road. It is also known as the Florence Station Bridge, after the railroad station once active here.

The structure is braced by a square steel bar bolted to the upper chords and anchored to the ground on the easterly side. Reconstructed in 1974, the timber deck now stands on four steel beams independent of the lattice truss; the truss supports only its own weight and snow load. The south abutment is faced with concrete and augmented by two piers set in the creek bank, one under each stringer. Notice the slate roof. Few of Vermont's covered bridges still have them.

The easiest way to get to the Cooley Bridge is to drive through the Depot Bridge. Drive 0.6 miles to Route 7, crossing the railroad track and bearing right. Turn right on Route 7 and turn right again on Bridge Street where it leaves Route 7 just north of the Country Store. Drive 1.2 miles to the bridge.

> ### Depot Bridge
>
> TOWN: Pittsford
> DATE: 1840
> TRUSS LENGTH: 120'6"
> BUILDER: Abraham Owen
> TRUSS TYPE: Plank-lattice

The Depot Bridge used to be an important connection between industrial Pittsford and the busy Florence Station. It spans the serpentine Otter Creek amid cornfields.

Cooley Bridge

Cooley Bridge

TOWN: Pittsford
DATE: 1849
TRUSS LENGTH: 50'3"
BUILDER: Nicholas Powers
TRUSS TYPE: Plank-lattice

The Cooley Bridge spans Furnace Brook in the rolling green countryside between the villages of Pittsford and Proctor. The first sight of the exaggerated overhang at the portals of the little bridge might remind you of a Conestoga wagon. A closer look will reveal a no-nonsense, practical application of Ithiel Town's plank-lattice truss.

The Cooley was built by Nicholas Powers using the same gable-end treatment found in other bridges in the Pittsford area. Still, the eight-foot gable-end overhang is a little out of proportion for a fifty-foot span. Perhaps Powers overdid the overhang on purpose—stories handed down tell of his droll sense of humor.

The Cooley Bridge is named for the family who owned the land around it. The family is descended from Gideon Cooley, veteran of the French and Indian Wars. He applied to his commanding officer, Captain Ephraim Doolittle, for a deed. Doolittle, anxious to settle the town, granted him the land in 1769 as a gift provided he worked it. With this three hundred acres on the west side of the

The Cooley Bridge, constructed early in the career of Nicholas Powers, one of Vermont's premier builders, features gable-end overhangs that are unusually long, even for the Pittsford area.

creek and one hundred acres deeded to his brother Benjamin on the east side, the tract was known as the Cooley Farm.

If you drive through the Cooley Bridge and proceed for 0.7 miles, you will come to Stevens Road and the Gorham Bridge. To return to Route 7, turn left on Stevens Road and drive 0.4 miles to Route 3. Turn left here and proceed 1.5 miles to Route 7.

Gorham Bridge

Builder Abe Owen's Gorham Bridge (the one Avery Billings lamented) is the southernmost of the Pittsford spans, crossing Otter Creek with its eastern portal facing the Pittsford-Proctor Road. The bridge is jointly owned and maintained by the towns of Pittsford and Proctor.

One hundred and fourteen feet long with the signature portal overhang, it stands on high abutments out of reach of the river's lesser freshets. Major repairs were made in 1956. Except for two distribution beams tie-bolted under the deck timbers, the bridge remains structurally the same as the builder left it.

Gorham Bridge

TOWN: Pittsford-Proctor
DATE: 1841
TRUSS LENGTH: 114'2"
BUILDER: Abe Owen/ Nicholas Powers
TRUSS TYPE: Plank-lattice

This is one of the bridges Nicholas Powers learned on. He was only twenty-four years old when he worked on it with Owen.

Twin Bridge

Rutland has replaced all of its covered bridges with modern spans. Only the Twin Bridge still exists, not as a bridge, but as a shed.

The story began in 1849, when the Rutland selectmen contracted with Nicholas Powers to replace a bridge spanning East Creek lost to high water. The new bridge was not in service very long before another flood came along. In contrary fashion, the East Creek dug itself an additional channel just beyond the new bridge, cutting off traffic once again.

There was nothing to be done but to bridge the new channel. Powers contracted to build another bridge a mere twenty feet from the first. The second bridge, a sixty-foot plank-lattice span and twin to the first, was opened to traffic in 1850.

Both bridges served until 1947, when the East Pittsford dam burst. One of the bridges was destroyed outright,

> **Twin Bridge**
>
> TOWN: Rutland
> DATE: 1850
> TRUSS LENGTH: 64'6"
> BUILDER: Nicholas Powers
> TRUSS TYPE: Plank-lattice

The Gorham Bridge spans a deceptively placid Otter Creek on a beautiful late fall day. Its construction sparked controversy among bridge- and tax-weary town residents.

Nicholas Powers' Twin Bridge is still a bridge in the structural sense, even though it has been five decades since it spanned water.

while the other survived to be dragged out of the creek bottom and converted into a storage shed.

The Twin Bridge remains a shed, and it is located near the cement bridge crossing East Creek on East Creek Road. Drive east on Prospect Hill Road where it leaves Route 7, 3.3 miles south of the junction with Route 3. When the road ends, turn south to the cement bridge. The shed is on the left, south of the bridge. Continue past the shed and you will come back to Route 7.

Brown Bridge

The 112-foot Brown Bridge stands over the Cold River, where crystal clear mountain water flows through boulders and stones on its way to Otter Creek. Brook and rainbow trout wait in the pools.

Standing in the shadow of hemlock and maple, surrounded with only the sounds of wind in the trees and the rushing of the stream, Nicholas Powers' last bridge is a fitting monument to one of Vermont's master builders.

The full length of the bridge is ventilated where the siding stops short of the eaves. The Brown Bridge's roof is still shingled with slate, one of the few original slate

> **Brown Bridge**
>
> TOWN: Shrewsbury
> DATE: 1880
> TRUSS LENGTH: 112'6"
> BUILDER: Nicholas Powers
> TRUSS TYPE: Plank-lattice

roofs left. Except for iron tie-rods between the trusses, the bridge structure remains unchanged. It is one of the finest and most original Town-lattice bridges in the state, says timber bridge engineer Gilbert Newbury. "It is straight, strong, and true, even with the heavy slate roof."

One of the abutments stands as testimony to Powers' Yankee sagacity. He avoided a considerable amount of stone work by placing the bridge so he could build a relatively small stone pad perched on an immense streamside boulder.

The Brown Bridge crosses the Cold River on Upper Cold River Road, just below the bend in the road to Shrewsbury. Visitors should be aware that the road is narrow and steep and is not maintained after the first snows. Cold River Road can be reached by turning east off Route 7 under the traffic light at North Clarendon, 1.1 miles south of the junction with U.S. Route 4.

CLARENDON

The land that became the Town of Clarendon was claimed under three different titles and became a flashpoint for the conflict between the Green Mountain Boys and New York officials. The first title, to Durham Township, was

The Brown Bridge, the last built by Nicholas Powers, is a fitting monument to one of Vermont's master bridge builders. The rushing waters of the Cold River are favored by trout fishermen.

granted in 1744 to Colonel John Henry Lydius of Albany, New York, by Governor Shirley of Massachusetts. New Hampshire Governor Benning Wentworth made the Clarendon grant in 1761. The Socialborough grant was issued by New York Governor Dunmore in 1771. Not surprisingly, when the various settlers came together, the fur started to fly.

After New York officers arrived to support the New York titles, Ethan Allen and one hundred Green Mountain Boys marched in to support the New Hampshire grants in 1773. Ultimately, through the persuasion of kangaroo courts backed by rifles and cutlasses, most of the inhabitants agreed to purchase New Hampshire titles.

In consequence of their actions, a New York law was passed adjudging the Green Mountain Boys to be guilty of felony and to suffer death without trial or benefit of clergy if they did not give themselves up. Allen dared them to come for him. The onset of the Revolutionary War resolved the standoff.

Kingsley Bridge

The Kingsley Bridge was built in 1836 by T. K. Norton, a Clarendon carpenter. The plank-lattice structure crosses Mill River surrounded by a stand of pine.

The bridge serves East Street just south of Rutland Municipal Airport. In 1825 Mr. and Mrs. John H. Kingsley operated a carding mill, a sawmill, and a gristmill near where the bridge now stands. Today, the bridge lies downstream from a falls and a mill.

The bridge was closed for repair in 1950 and closed again for restoration in 1985. It was reopened in May 1987. There are steel cables on all four corners to give lateral reinforcement, but the basic structure remains unchanged from the original.

The road is narrow and lined with no parking and no trespassing signs, making it difficult to stop to visit T. K. Norton's legacy. The bridge is most easily found by leaving Route 7 east on Gorge Road, 3.4 miles south of the traffic light at North Clarendon. The author has tried to get to the Kingsley Bridge directly from the Brown Bridge several times, but has managed to get lost nearly every time.

Kingsley Bridge

TOWN: East Clarendon
DATE: 1836
TRUSS LENGTH: 120'6"
BUILDER: T. K. Norton
TRUSS TYPE: Plank-lattice

The Kingsley Bridge is named for a family that operated a nearby mill. Unfortunately, visitors should expect a cold welcome—the mill and the narrow road the bridge serves are posted with no trespassing signs.

Sources

Allen, Richard Sanders. *Covered Bridges of the Northeast.* Brattleboro, VT: Stephen Greene Press, 1957.

Brandon Vermont: A History of the Town 1761-1961. Brandon, VT: Town of Brandon, 1961

Congdon, Herbert Wheaton. *The Covered Bridge.* Middlebury, VT: Vermont Books, 1973.

Hemenway, Abby Maria. *The Vermont Historical Gazetteer, Vol. III.* Burlington, VT: A. M. Hemingway: 1877.

Petersen, Max B. *Salisbury: From Birth To Bicentennial.* South Burlington, VT: The Offset House, 1976.

Vermont Agency of Transportation Covered Bridge Study. Prepared for the State of Vermont Agency of Transportation by McFarland-Johnson, Inc., Binghamton, New York, 1995.

Vermont Historical Society. "Merrily The Search Goes On: About Vermont Place Names" *News and Notes* (monthly newsletter), Vol. 2, Number 10. (June 1951)

Vermont Historical Society. "Merrily The Search Goes On: About Vermont Place Names" *News and Notes*, Vol. 3, Number 1. (July 1951)

Vermont Historical Society. "'Lemon Fair' and 'Gulf': The Trail Grows Warmer." *News and Notes*, Vol. 3, Number 5. (January 1952)

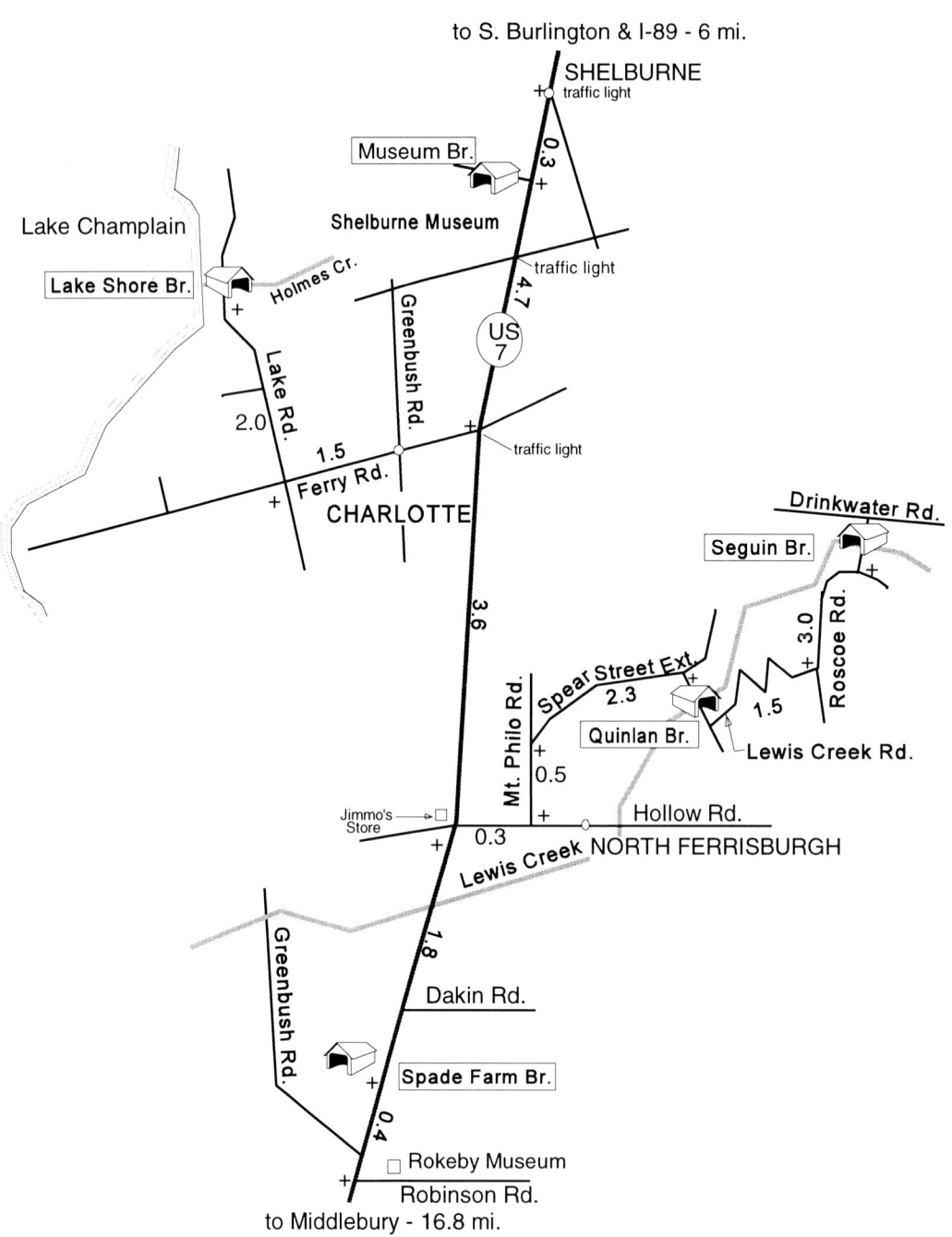

Tour 3
+ = *Distance marker*
Distances in miles (estimated)
Map not to scale
© *1996 Joseph C. Nelson*

The WOODEN BRIDGES of CHARLOTTE

Tour 3

There are five covered bridges in Charlotte and the surrounding area. While other areas of the state may have more bridges, the unique character of each of the Charlotte bridges makes them well worth viewing.

Three of Vermont's last nine Burr-arch bridges are found here. One of the three, the Museum Bridge, is one of the last two double-barrel arch bridges in the state and one of the last seven in the nation. The Lake Shore Bridge is not only one of just three surviving tied-arch truss spans in Vermont, it also stands at the lowest elevation in the state. The Spade Farm Bridge is possibly one of the two oldest in the state using the lattice truss.

CHARLOTTE

Charlotte was granted to Benjamin Ferris and sixty-four others by Benning Wentworth. The proprietors hailed from Connecticut and New York, and the town was not settled until after the Revolutionary War. When the first settlers arrived, they found the land particularly well suited to agriculture, and the town flourished from the start.

Charlotte, by the way, is pronounced Shar-lot, with the accent on the second syllable. If you are visiting Vermont, it is hard to get used to the different pronunciation. You will not be the only non-native in the area, though—two of the five covered bridges are also "from away."

Museum Bridge

The Museum Bridge stands over Burr Pond a few yards from the west side of Route 7 in Shelburne Village. Built by master builder Farewell Wetherby in 1845, the bridge served the town of Cambridge, forty-five miles to the north of the bridge's current location, for more than a century.

When the state made plans to replace the Cambridge double-laner, Mrs. J. Watson Webb asked the highway department to donate the old arch bridge to the Shelburne Museum. It has served as an entrance to the museum since 1951.

The state agreed to donate the bridge if the museum moved it. Fortunately, the museum already had a good deal of experience in finding historic structures and bringing them to the museum grounds. R. V. Milbank, professor of Civil Engineering at the University of Vermont, and W. B. Hill and Company of Tilton, New Hampshire, helped the museum disassemble the 169-foot bridge, move it by truck, and reassemble it in its current location. There was only one problem—a bridge needs water, so a pond had to be dug to make the restoration complete.

The museum has placed an exhibit of wagons and coaches complete with plaster horses in harness inside the bridge to create a period atmosphere and appearance. There are also historic buildings surrounding the bridge's location on the museum grounds. The setting and coach display provide a rare look at how covered bridges served the users for which they were designed.

Technically, the Museum Bridge is not a true Burr-arch bridge, but a bridge built in the "manner" of the Burr patent. The main departure is in the use of the multiple-kingpost truss instead of Burr's truss where the vertical posts are supported by crossed braces rather than the simple single brace. The Structures Division of the Agency of Transportation describes all of the state's Burr bridges as "multiple kingpost plus Burr." True to Burr's usage, the ends of the museum bridge's segmented timber arches are butted into the supporting piers below the main stringers, a practice not followed by all of Vermont's "Burr"

Museum Bridge

Town: Shelburne
Date: 1845
Truss length: 156'
Builder: Farewell Wetherby
Truss type: Burr arch

When the state made plans to replace the double-lane Cambridge "Big Bridge," Mrs. J. Watson Webb asked the highway department to donate the old bridge to the Shelburne Museum. It has served as an entrance to the museum since 1951.

bridge builders. Many of them terminate the arches at the ends of the lower chords.

The Burr timber arch is a neat idea, based on the same principle as that of a stone arch—each piece transmits the weight of the structure to the other, and ultimately to the banks of the stream. Each arch segment, rabbeted on both ends, is fitted one into the next on both sides of the multiple-kingpost truss and bolted together through the vertical truss members. The arrangement of the Museum Bridge is very like that of the Pulp Mill Bridge at Middlebury-Weybridge, except there the paired timber segments were replaced with single laminated bows.

Lake Shore Bridge

The Lake Shore Bridge crosses Holmes Creek just a few feet from Lake Champlain. It stands in a grove of rugged old willows, at the lowest elevation of all of the covered bridges in Vermont.

Leonard Sherman built the forty-foot span in the late 1890s using the tied-arch truss. The Lake Shore Bridge, also known as the Holmes Creek Bridge, is one of the last three surviving tied-arch bridges in the state. The other two are Bests Bridge and Bowers Bridge in Brownsville.

The construction of the Lake Shore span is unique in that the tied arch is used in conjunction with what appears to be a kingpost truss—the main stringers on each side of the bridge are supported with three vertical beams bolted to each arch. The Brownsville bridges' arches support the main stringers with iron rods.

An Agency of Transportation inspection team found the bridge in trouble in the fall of 1993 with "considerable movement of the superstructure under normal traffic loading." The team concluded that the kingpost/tied-

> **Lake Shore Bridge**
>
> TOWN: Charlotte
> DATE: 1898
> TRUSS LENGTH: 40'2"
> BUILDER: Leonard Sherman
> TRUSS TYPE: Tied arch

The Lake Shore Bridge is protected from the rigors of lakeside wind and weather by a stand of rugged old willow trees.

> **Charlotte's Little Burr Bridges**
>
> The unknown builders of the Quinlan and Seguin bridges were singularly talented individuals—similarites in the two structures indicate that it is likely both bridges were built by the same people. Very sophisticated and intricate construction details make these two among the best built wooden bridges in the state. Intricate "keys," easily visible in the Seguin Bridge, are used in the tension connections in the bottom chords. The elaborate roof framing systems include "birds-mouth" notches in the rafters and beams. Also, the top chord lateral braces are set in an elliptical shape, which is very difficult to do and not copied elsewhere.

arch combination was not the original construction. The tied arch, a lamination of six two-inch by six-inch planks, appears to have been mortised into the bottom chords at the time of the original build. The kingpost diagonal beams, probably added later, do not contribute to the support of the bridge stringers.

The Lake Shore Bridge was rehabilitated in 1993 by Milton Graton Associates. Additional work was done in 1994 by Paul Ide and Jan Lewandoski. In the interest of preservation, and because of repeated damage caused by overweight vehicles, the AOT has recommended the construction of a permanent bypass next to the covered bridge.

To find the Lake Shore Bridge, drive five miles south on Route 7 from Shelburne Village, then turn right on Ferry Road at the intersection at the traffic light. Follow the signs to the ferry landing for 1.5 miles, then turn right on Lake Road. Follow Lake Road two miles to the bridge. A town beach and parking lot are nearby.

Quinlan Bridge

The Quinlan Bridge crosses Lewis Creek in the midst of the rolling farmland that was once the bottom of the ancient Champlain Sea. To the west, Mount Philo jags upward—Lake Champlain and the Adirondacks lie beyond. To the east, the foothills rise to the Green Mountains.

The eighty-six-foot multiple-kingpost Burr-arch bridge, built in 1849, stands beside a tall hickory tree and a farmhouse dated 1798. The span has been strength-

> **Quinlan Bridge**
>
> TOWN: Charlotte
> DATE: 1849
> TRUSS LENGTH: 85'6"
> BUILDER: Unknown
> TRUSS TYPE: Burr arch

The Quinlan Bridge was built next to Sherman's sawmill using a multiple-kingpost truss with Burr arches. The bridge stands in a valley that was once under the ancient Champlain Sea.

ened with two steel beams installed under the existing floor system, probably in 1949 or 1950. The deck is additionally supported by a bolster beam added to the underside of the deck beams. The original stone abutments have been capped and faced with concrete.

The Quinlan Bridge is named for a farm family that had extensive holdings in the Lewis Creek valley. The road crossed the creek at a ford at one time, but the ford was thought to be a poor place to maintain a bridge, so the crossing was moved upstream next to Sherman's sawmill. The span is also sometimes referred to as the Sherman Bridge.

To find the Quinlan, turn left off Route 7 onto Hollow Road in North Ferrisburgh. The turn is marked by Jimmo's Store opposite. Take the first left, Spear Street Extension, then bear right at a fork in the road. Drive about 2.3 miles to the bridge.

Seguin Bridge

Drive through the Quinlan bridge and turn left on Creek Road. Continue until the road ends at the intersection of Rosco Road. Turn left and drive about three

> **Seguin Bridge**
>
> TOWN: Charlotte
> DATE: 1849
> TRUSS LENGTH: 70'3"
> BUILDER: Unknown
> TRUSS TYPE: Burr arch

miles to the Seguin Bridge, also called the Upper Bridge. Look for the waterfalls upstream. The countryside here is mixed heavy second-growth woods and abandoned pasture and crop land. Go slow—the road is a curvy one as it follows the course of the winding stream.

The seventy-foot Seguin Bridge uses the multiple-kingpost Burr-arch truss, and, like the Quinlan, was built in 1849 to cross Lewis Creek. The portals are decorated with simple pilasters.

Extensive repairs were performed in 1949, leaving the bridge unaltered from the original builder's plan. Authentic repair work was done again in the fall of 1994 by Paul Ide and Jan Lewandoski. The work included replacement of portions of the bottom chords, several vertical posts, and a set of bearing blocks.

In pursuit of the state's policy of preservation, the Agency of Transportation has recommended the construction of a permanent bypass rather than a reconstruction of the covered bridge with "non-traditional" materials and techniques. The recommendation was made in consideration of the long detour necessary for overweight vehicles to avoid the crossing.

The Seguin Bridge is considered one of the best constructed in the state. It has supported traffic without major reconstruction for more than 150 years.

Justin Miller's bridge came to the Spade Farm from Lewis Creek in 1958, seven years after the Museum Bridge came to Shelburne. It served as an attraction for a popular restaurant that once did business here. Bridge historian Richard Sanders Allen dated this bridge circa 1850.

Spade Farm Bridge

The Spade Farm Bridge stands on the west side of Route 7 in front of the Old Farm Restaurant. It was moved here in 1958 from where it crossed Lewis Creek in North Ferrisburgh. Justin Miller built the eighty-six-foot plank-lattice structure in 1824.

The old bridge has fallen into disrepair over the years. Viewers should be aware that the bridge is on private property and that the floor is dangerously rotted in many places. Finally, a word to the wise: watch out for the geese.

Spade Farm Bridge

TOWN: No. Ferrisburgh
DATE: 1824
TRUSS LENGTH: 85'6"
BUILDER: Justin Miller
TRUSS TYPE: Plank-lattice

Sources

Allen, Richard Sanders. *Covered Bridges of the Northeast.* Brattleboro, VT: Stephen Greene Press, 1957.

Vermont Agency of Transportation Covered Bridge Study. Prepared for the State of Vermont Agency of Transportation by McFarland-Johnson, Inc., Binghamton, New York, 1995.

Ziegler, Philip. *Sentinels of Time: Vermont's Covered Bridges.* Camden, ME: Down East Books, 1983

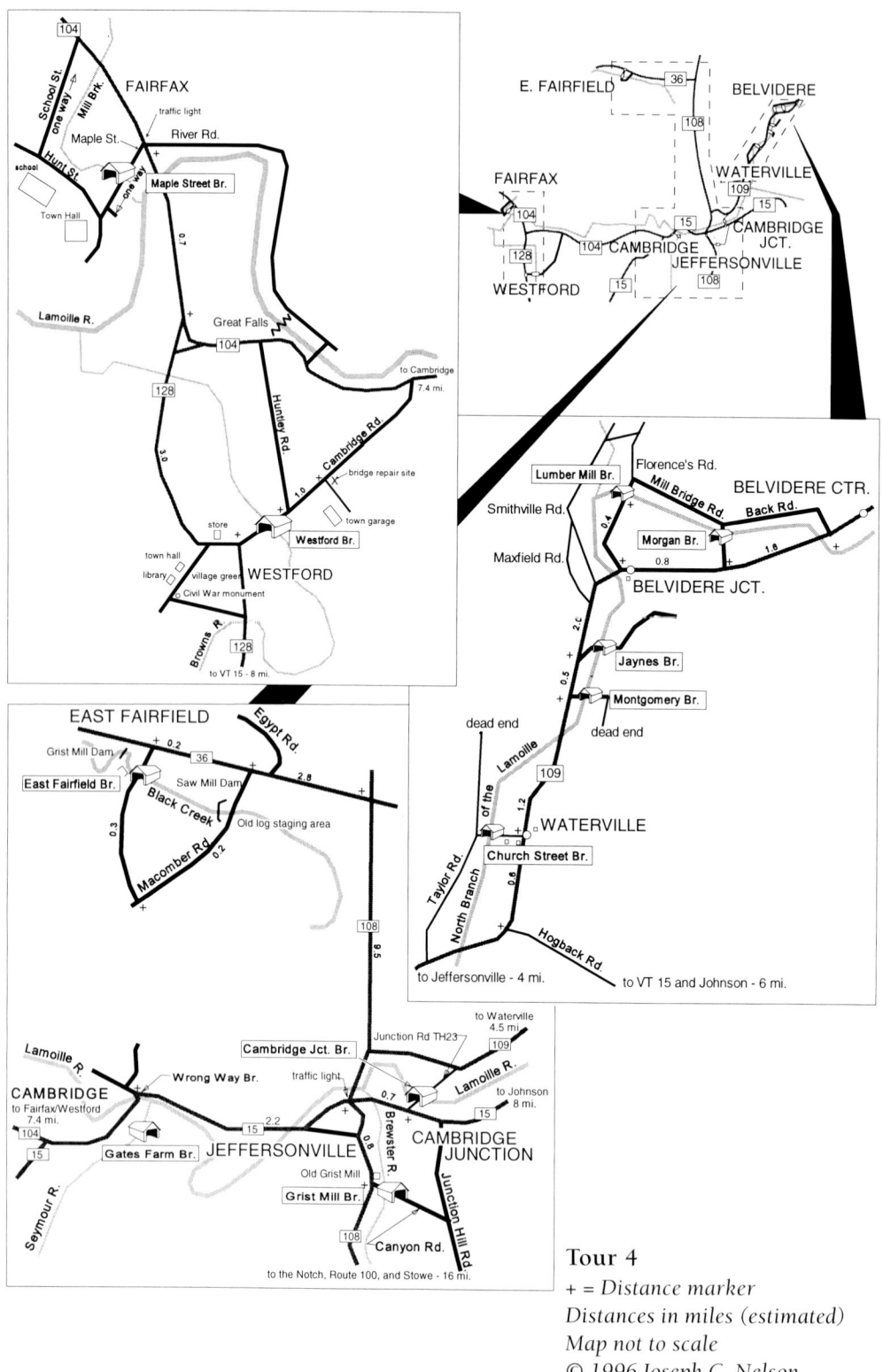

Tour 4
+ = Distance marker
Distances in miles (estimated)
Map not to scale
© 1996 Joseph C. Nelson

The LAMOILLE RIVER and the NORTH BRANCH

Tour 4

The northern Green Mountain watershed feeds an extensive system of rivers and streams with snow-melt in the spring and with summer and fall rains. The Lamoille River rises in Greensboro and loops and curls its way roughly seventy miles—as the crow flies—to Lake Champlain. The North Branch flows into the Lamoille some thirteen miles from its source in Eden. Browns River curves down from Mount Mansfield and through twenty-two miles of rich bottom land. Black Creek flows north to the Missisquoi River.

People came to these lush valleys starting in the late 1700s, and soon settlements grew, dams rose, mill wheels turned, roads were laid out, and bridges replaced fords. The towns and villages of Westford, Fairfax, Cambridge, Fairfield, Waterville, and Belvidere came to be.

Eleven covered bridges remain from the days of growth in this region. The oldest known surviving bridge dates back to 1835, while the youngest was completed in 1897. Four are arch bridges, and the rest are queenpost.

Westford Bridge

Voters in Westford resolved in 1836 to build a single arch bridge over Browns River for a cost not to exceed six hundred dollars. The bridge was originally part of the Vermont Market Road ordered built by the Vermont Supreme Court in 1827. Workers finished construction of the bridge in 1838, but the road project as a whole was never completed.

The ninety-seven-foot span was in service until a concrete and steel bridge replaced it in 1965. After 127 years of continuous use, the old bridge could no longer handle modern traffic. In 1976, after years of abandonment, the townspeople and a reserve Seabee battalion from Burlington repaired the bridge for the National Bicentennial. The repair work did not include fixing the wooden arches, which had rotted where they met the stone abutments, so the old bridge sagged noticeably in the years that followed. It was closed to foot traffic in May 1987.

In February 1987 Westford resident Tom Kennedy called a special meeting to save the bridge. The Westford Historical Society was formed, and it ultimately selected Graton Associates of Ashland, New Hampshire, to do the restoration. The Gratons advised that the bridge be removed from the river before winter as it was in danger of collapse.

Work began in October 1987. Graton raised the bridge on timber cribbing and built false work under it. A team of oxen pulled the bridge off the river on log rollers. Graton then used his tractor-trailer truck and the town's bucket loader to move the bridge uphill to the town garage property while residents stood in rain and mud to watch the work. The National Geographic Society filmed the operation for a documentary on Milton Graton's life as a bridge restorer. Also filming the event was a crew from the America-How-Are-You TV series.

The bridge was restored where it stood next to the town garage, but it remained there for the next several years—the project went on hold, waiting for funds. In 1995 the Westford Historical Society received a $36,000 grant from the Agency of Transportation through the federal Intermodal Surface Transportation Enhancement Act (ISTEA) to complete the project by replacing an abutment, moving the bridge back over the river, and landscaping the area.

The grant paid for 80 percent of the completion costs. The balance of the funding was raised by the Historical Society. The society worked four years to raise money to restore the bridge without cost to the town government—the moving of the bridge and the restoration, the first phase of the project, cost $50,000. Funding came from

> **Westford Bridge**
>
> TOWN: Westford
> DATE: 1838
> TRUSS LENGTH: 96'6"
> BUILDER: Unknown
> TRUSS TYPE: Burr arch

The Westford Bridge was built in 1838 to serve the never completed Vermont Market Road. Bypassed by a concrete bridge in 1965, the bridge deteriorated and was in danger of collapse before the community came to the rescue in 1987 and hired Milton Graton to restore it.

community fairs, ham suppers, and grants. Asked what drove her and the townspeople to make such an effort, Historic Society President Caroline Brown replied, "It must be pride of place. The bridge is one of Westford's few remaining treasures."

The bridge was scheduled to be back in its proper place at last in the late 1990s, so hopefully the final phase of the restoration will take place. With the camber or curve of the span restored, the side boards will be trimmed and the three-inch floor planking installed. The bridge will serve as a foot bridge and park.

Upon replacement, the Westford Bridge will be found in Westford Village just off Route 128 where it crosses Browns River on Cambridge Road. A general store stands at the north end of the village green. Cambridge Road joins Route 128 east of the store.

Fairfax

Fairfax had its beginnings in 1763 as a plantation granted to Edward Burling and sixty-four patentees by Benning Wentworth, Governor of New Hampshire in the name of King George III. None of the original sixty-four patentees ever settled in Fairfax.

Fairfax-by-the-Lamoille was richly endowed with fertile plains and an abundance of water power, but it was wilderness. Travel between settlements was difficult much of the year, so each growing village had to be self-sufficient. An enterprising settler built a fulling mill on Great Brook. After that came a tannery, sawmills, a gristmill, a starch factory, a chair factory, and a carriage shop scattered along a nine-mile stretch of river. With all of this industry, Great Brook came to be called Mill Brook.

Development also began early on the Lamoille River near Great Falls, said to be "considered the most valuable source of water power in Vermont." A cloth mill was built here in 1824. A gristmill started business in 1850 and a woolen mill opened in 1864. An electric power station and dam came in 1903.

A lively competition for skills existed between settlements. In 1826 the town fathers offered a harness maker from a neighboring town the position of toll gatherer for the Lamoille River Bridge serving what is now Route 104. With the job of collecting cash tolls came a rent-free house and plenty of time to make harness.

As in many of Vermont's river towns, the mills and bridges in Fairfax are vulnerable to flooding. The first bridge over the Lamoille at Fairfax, built in 1792, was destroyed by a spring freshet in 1814, and the spring freshet of 1832 took its successor. The two-lane arch bridge replacement was lost to the massive flood of 1927, which devastated the town. It carried the several mills and the double bridge at Fairfax Falls over the Great Falls, and, in all, four covered bridges in town washed off their abutments. Only the Maple Street Bridge was salvaged and reset—the others were replaced with concrete and steel.

Maple Street Bridge

The Fairfax town history says that "the covered cross-x bridge over Mill Brook on Maple Street in the village was built in 1865." The terms "cross-x bridge" and "ex bridge" are used in town records to describe bridges built "in the manner" of Ithiel Town's patented plank-lattice truss. With an inside clearance of seventeen feet, the bridge is one of the widest in the state, probably to accommodate two-way village traffic. The Vermont Divi-

---

Maple Street Bridge

TOWN: Fairfax
DATE: 1865
TRUSS LENGTH: 56'6"
BUILDER: Kingsbury and Stone
TRUSS TYPE: Plank-lattice

With an inside clearance of seventeen feet, the Maple Street Bridge is one of the widest in the state. After being brought back to its abutments after the 1927 flood, it was thought by some that the bridge was reinstalled with its original east end facing west, which makes it seem to lean.

sion of Historic Sites identifies the builders as Kingsbury and Stone.

Town historians recall that when the fifty-seven-foot bridge was returned to its abutments after being washed away during the 1927 flood, ". . . it was replaced with its east end facing west. Because of the error, they say, the bridge seems to be leaning." This last, the historian writes, "is disputed by some."

Major work was done in 1975, when the stone abutments were faced with concrete and a metal roof was laid over the wooden shingles. In 1990 the bridge underwent timber restoration by Jan Lewandoski.

The bridge was sagging noticeably. It had extensive areas of rotten chord and lattice due to poor approach drainage and roof leakage. Repairs had been made by bolting in short four- to eight-foot pieces of plank.

Restoring the bridge would have resulted in the replacement of a large number of lattice planks merely because of damage to a few inches of their lower ends. Rather, sistering—slipping in and pinning new lattice next to the damaged lattice—was chosen as a method to both repair and add rigidity to the truss by increasing the number of double-pinned lattice-to-lattice and lattice-to-chord crossings.

The final appearance of the bridge was slightly changed by the presence of several lattice sisters, but the repairs were carried out within the structural system with the types of materials from which the bridge was originally built. The Vermont Division for Historic Preservation felt that the sistering alternative allowed the maximum amount of historic material to remain in the bridge, while satisfying the Agency of Transportation's desire for increased strength. The work was completed in 1990.

To find the Maple Street Bridge, enter Fairfax Village on Route 104 from the south. Turn left onto Maple Street at the village's only traffic light. Maple Street is one way, so after crossing the bridge, continue to Hunt Street and turn right, pass the school building, and turn right onto one-way School Street to return to Route 104.

Cambridge

There are three principal villages in the Town of Cambridge—Cambridge Center, Jeffersonville, and Cambridge Junction—and each prizes its own covered bridge. The Gates Farm Bridge is in Cambridge Center, the Grist Mill Bridge is in Jeffersonville, and the Cambridge Junction Bridge crosses the Lamoille River in Cambridge Junction. All of them are "arch" bridges based on the truss patented in 1817 by Theodore Burr.

Big Bridge and the Little Bridge

Cambridge once had two covered bridges: the Big Bridge, a 156-foot two-lane or double bridge over the Lamoille River, and the Little Bridge, an eighty-two-foot single span over the Seymour River, both serving the road that became Route 15. Long-time residents of the area recall the days when the village green extended from Little Bridge to Big Bridge along Route 15 and then to the banks of the Lamoille. All the big doings, like the Fourth of July celebration, were held there.

Farewell Wetherby built the double bridge in 1845, assembling the trusses in a field owned by the Gates family close to the chosen site. He erected the span over the

The Tale of the Wrong-way Bridge

The concrete and steel span that replaced Cambridge's old double-barrel covered bridge was built with a decided curve. It became known locally as the Wrong-way Bridge, because the north end of the structure points toward the village of Fletcher, where hardly anyone goes, forcing Jeffersonville-bound drivers to execute a hard right turn.

There are many stories about the replacement bridge crossing the Lamoille River at Cambridge. One states that the designers of the bridge specified that the steel girders were to be formed to accommodate a curving roadbed. Unfortunately, an error was made, either in the specifications or in the interpretation of them. The structure was delivered with the required bend in the wrong direction, and the lay of the road had to be changed to compensate for the goof.

Roger Boozan, a Fletcher native, refutes that story—he remembers that the bridge was built correctly. The plan was to build the curve so that a new road could run straight off the end of the bridge, up the hill, and on to Jeffersonville, avoiding the stretch of road beside the Lamoille River that was under water almost every spring. The new road was never built, and drivers have been tangling on the severe curve at the Pumpkin Harbor end of the bridge ever since.

Mr. Boozan's story dovetails with another oft-repeated anecdote. A road built straight off the end of the bridge and up the hill would head directly for St. Albans, using an abandoned railroad right-of-way. As it happens, St. Albans was the home of the highway superintendent at that time! The new road was not built, it is said, because the superintendent died before the work was done.

river the following winter on false work supported by the ice. An attached footbridge was added later. George Washington Holmes built the Seymour River span in 1897. Both builders employed a truss similar to that invented by Theodore Burr. The bridges have the signature segmented timber arch, but it is bolted to a multiple-kingpost truss instead of using Burr's counter-braced system.

In 1950 the state bypassed the double bridge with a new span of concrete and steel. The project also diverted the Seymour River away from the highway, saving the State Highway Department the cost of another new bridge and leaving Little Bridge unemployed. The new river channel ran through a farm owned by Earl Gates, closing access to sixty acres of his land. Access was restored when the state moved the Little Bridge from its old crossing place to the farm. The Highway Department gave the double bridge to the Shelburne Museum with the stipulation that the museum arrange to move it.

The Gates Farm Bridge stands in a cornfield beside Route 15. The bridge was originally built as the Little Bridge over the Seymour River by George Washington Holmes in 1897, but it was moved in 1950 after the river was diverted through the farm.

Gates Farm Bridge

For years, the Little Bridge could be seen standing in the field to the east of Route 15 near Wrong-way Bridge (see above). The Little Bridge on Gates Farm came to be known as Gates Farm Bridge. It fell on hard times and needed maintenance desperately. The state told Rex Gates, Earl's son, that the bridge was his to maintain. Gates retorted that the state made a mistake when they diverted the Seymour through his land. The state created the problem, so the state must maintain the bridge[1].

As the old truss began to sag, the State Highway Department (now the Agency of Transportation) put supporting posts under the stringers each spring. Before winter, the posts were removed and two steel beams were

---

[1] Earl Gates had an agreement written declaring the state would install and maintain the bridge until "other access is made available." The agreement was signed by he and his wife and recorded by the town clerk, but it was not signed by the state. Here began legal interpretation and delay until the bridge was ready to collapse. The state funded the bridge restoration.

laid on either side of the roadway and tie-bolted to the stringers below. The state made the swap twice a year because during growing season the farmer needed the full width of the roadway for his equipment, and the posts were vulnerable to ice and flotsam in winter and spring.

After years of seasonal submergings in high water, the ninety-seven-year-old structure failed. The old bridge was lifted off its abutments in the fall of 1994 and disassembled. The plan called for all structurally sound members to be saved, but damage was so extensive that 80 percent of the bridge had to be replaced. The work was done by Blow and Cote of Morrisville. Agency of Transportation engineer Gilbert Newbury, trained in timber engineering, designed the replica bridge for the Maintenance Division of the Agency of Transportation.

Because the bridge stands in a flood plain, it was necessary to raise the bridge structure out of reach of high water and floating ice. Newbury couldn't increase the height of the abutments because that would steepen the slopes in and out of the bridge, making it difficult for the farmer to get his equipment through. He couldn't build up the road because the U.S. Corps of Engineers frowns on bringing fill into a flood plain. The problem was solved when Newbury raised the bridge on eighteen-inch white oak bolster timbers and hung new floor beams under the chords by steel rods. This raised the trusses eighteen inches higher over the river, yet the new bridge floor remains at the top edge of the original abutment backwall as before. While the floor system is still exposed to high water, the floor is both cheaper and easier to replace than the bridge trusses. Newbury attributes the floor suspension idea to the Howe Truss used in the Rutland Railroad Bridge in Shoreham.

The renewed Gates Farm Bridge can be seen standing on the old abutments in the cornfield on the east side of Route 15, just before you cross the Lamoille River on the Wrong-way Bridge.

Gates Farm Bridge
TOWN: Cambridge
DATE: 1897
TRUSS LENGTH: 82'
BUILDER: G. W. Holmes
TRUSS TYPE: Burr arch

Grist Mill Bridge

Little is known about the eighty-eight-foot bridge spanning the Brewster River a few hundred feet upstream from the old gristmill. The bridge serving Canyon Road off

Route 108 south of Jeffersonville has been called the Grist Mill Bridge, the Scott Bridge, the Bryant Bridge, the Grand Canyon Bridge, and the Brewster River Bridge. No one knows who built it or when—oddly, there are no public records testifying to the events that lead to the building of a bridge. The town records concerning these events could have been lost in flood or fire, but no newspaper accounts have come down through the years either.

Except for the addition of two wooden joists tie-bolted under the roadway, the structure remains as it was when built. There is ready access to the stream bed here and the shoreline is open, providing easy viewing of the bridge from below.

The abutments are of interest in that they are of the original rubble stone laid up dry. Bridge builders used rubble stone, stone slabs, or cut stone to build abutments, and the original masonry of the east abutment can still be seen here. The west abutment was thought to be failing, so it was faced with concrete.

There is disagreement about how to fix failing unmortared stone abutments. Some modern engineers choose to

> **Grist Mill Bridge**
>
> TOWN: Cambridge
> DATE: Unknown
> TRUSS LENGTH: 87'7"
> BUILDER: Unknown
> TRUSS TYPE: Burr arch

The eighty-eight-foot Grist Mill Bridge crosses the Brewster River gorge. The span has been known as the Brewster River Bridge, the Bryant Bridge, the Scott Bridge, and fancifully, as the Grand Canyon Bridge.

fill voids or seams in the face stones with mortar—a practice called pointing—or to pour a thin reinforced concrete "face" over the surface of the original stone. Other engineers disagree with these techniques, however. They feel that the original concept of unmortared stone construction has the advantage of good drainage—the abutments do not trap water migrating through the stone to the river. They contend that mortar pointing and concrete facing traps water in the abutment and increases the chance of ice damage. Their solution is to fill voids and replace loose or broken stone with new stone wherever possible.

Because of the long detour needed to avoid the Grist Mill Bridge, and because snowplows using the structure far exceed its capacity, the Agency Of Transportation recommended that either the bridge be closed and a permanent bypass built, or that the bridge be rehabilitated to support more weight.

To find the bridge from Route 15, drive to the junction of Routes 15 and 108 at the traffic light by Gates' lumber yard and turn south. Follow the village street to the intersection with a traffic light island. Bear left, continuing south on Route 108. The Grist Mill Bridge stands on the east side of the road. Turn left onto Canyon Road about one hundred yards past the mill.

Cambridge Junction Bridge

The 140-foot Burr-arch structure spanning the Lamoille River at Cambridge Junction is known as the Cambridge Junction Bridge, the Junction Bridge, the Station Bridge, and the Poland Bridge. Built in 1887 by George Washington Holmes, Jason French, and Roscoe Fuller, it once gave the towns west of the Lamoille River access to the railway station that contributed its name to the place. The abandoned St. Johnsbury and Lamoille County Railroad tracks still pass a few yards from the south portal.

The bridge is located off Route 15 about a mile from the junction of Route 15 and Route 108 in Jeffersonville, connecting Routes 15 and 109. Until recently it was in heavy use by truck and car traffic alike, despite a sign saying, "No Trucks Allowed."

In March 1993 the Agency of Transportation inspected the bridge and found it to be "seriously distressed . . .

Cambridge Junction Bridge

TOWN: Cambridge
DATE: 1887
TRUSS LENGTH: 140'
BUILDER: G. W. Holmes
TRUSS TYPE: Burr arch

George Washington Holmes built the Cambridge Junction Bridge, which gave the communities west of the Lamoille River access to the railroad station.

the truss significantly sagged," so the bridge was closed. The agency warned that stabilization of the structure was necessary to avoid failure from self-weight and snow load. If vehicles were to use the bridge again, reconstruction using nontraditional materials and methods would be required.

The report notes that this bridge is the second longest single span in Vermont after the plank-lattice Bartonsville Bridge. "It is possible [the Cambridge Junction Bridge] is the longest Burr Arch in the U.S."

The bridge very nearly was not built. The villages of Belvidere and Waterville were in favor of a bridge over the Lamoille to shorten the travel distance to the rail junction at Cambridge. Cambridge voters, however, could see no advantage to themselves and blocked the expenditure of tax dollars to build the bridge.

The impasse dragged on until Luke P. Poland of Waterville took up the cause. Poland, a lawyer, had served as Chief Justice of the Vermont Supreme Court, Representative to the Legislature, then Senator from Vermont in Washington D.C. before retiring from office in 1884 and returning to Waterville. Judge Poland led Belvidere and Waterville residents in bringing suit against the town of Cambridge at the Lamoille County Court in 1885. A commission ruled in favor of the petitioners, and the court ordered the bridge built. The Cambridge voters stalled compliance until the spring of 1887. Poland did not live to use the bridge, dying in his hayfield the following July, but his name remains associated with the bridge to this day.

The old span had not been closed very long in the 1990s before concerned citizens applied for funds to set it on new abutments bridged with steel beams. The grant application is supported by the Cambridge and Waterville selectboards, the local chamber of commerce, and the Lamoille County Planning Commission. Some people may bemoan having to wait to cross a single-lane bridge, but few of them want to see an old covered bridge die.

East Fairfield

East Fairfield lies thirteen miles north of Jeffersonville by way of Routes 108 and 36. The Town of Fairfield was founded in 1763, when New Hampshire Governor Benning Wentworth granted the land to petitioners in Fairfield, Connecticut. The Governor of New York granted the same land to another group of people in 1774. The first settler from Connecticut cleared land in 1787, and industry grew along the banks of Black Creek and the Fairfield River.

East Fairfield Bridge

The East Fairfield Bridge is a small queenpost span that crosses a mill pond created by a dam built to power a gristmill. The mill building foundations can be seen at the south end of the bridge. The dam, sluice, and foundations of a sawmill are visible just upstream. Above the sawmill the old staging area can still be seen next to the creek. Logs were brought here by teams of horses, then rolled into the creek. The logs were pulled into the mill by drag chain. Toward the Bakersfield town line there was a tub factory—where butter tubs, sugar tubs, and watering troughs were made—a fulling mill, tanneries, and a brick works, all of which are gone without a trace.

East Fairfield Bridge

TOWN: Fairfield
DATE: 1865
TRUSS LENGTH: 68'
BUILDER: Unknown
TRUSS TYPE: Queenpost

The East Fairfield Bridge now stands quiet and isolated over a small mill pond. The neighborhood was once very busy, alive with the clangor of water driven mills. The viewer can look upstream to see the old sawmill dam and downstream to the gristmill dam.

The town has worked to keep the bridge in use. Repairs were made in the early 1940s, and the span was reconstructed in 1967. In the winter of 1973–74, using the ice to support staging, selectmen Bernard Conner and Francis Howrigan and Howrigan's son Michael did some much needed repair work. The ends of the truss braces supporting the queenposts had decayed, and the north side of the bridge needed to be jacked up and strengthened. Truss rods were put through the frame of the bridge to prevent it from spreading. Despite all of the effort, the bridge was closed to traffic in 1987, and its condition deteriorated rapidly. Work on the floor in the early 1990s has allowed the bridge to be reopened for foot traffic, but vehicles are blocked from entering by posts barring the portals.

The Agency of Transportation Covered Bridge Study team inspected the bridge in May 1995. The team advised that to preserve the bridge for the future the structure should remain closed to vehicular traffic and that the bridge requires "extensive" stabilization measures to retain its capacity to support its self weight and snow loading.

Find the East Fairfield Bridge by driving three miles east on Route 36 from the intersection of Route 108. As you enter East Fairfield Village, pass Egypt Road on the right and Macomber Road on the left. Turn left 0.2 miles past Macomber Road. The bridge crosses Black Creek at the bottom of a small hill.

WATERVILLE AND BELVIDERE

The covered bridges of Waterville and Belvidere seem to share a kinship. All of them span the same stretch of the North Branch of the Lamoille River, all of them were built in mature communities, likely replacing older bridges, and they were probably built by local carpenter-members of the two communities using the same resources and stores of knowledge. All five of the Waterville-Belvidere bridges use the queenpost truss, a truss the carpenters knew well, as it is often used in barns.

The descendants of the craftsmen who built the bridges, showing more than a little faith in the work-

---

How Egypt Road Got Its Name

Just north of the covered bridge, Egypt Road leaves the village of East Fairfield and winds to the north. Egypt is a curious name for a road in northern Vermont, so not surprisingly, there is a curious story to go with it.

The upper counties of Vermont had a terrible growing year in 1816, which became known as the year with no summer. Monthly frosts destroyed crops until there was not much left to harvest, except on Nathaniel Foster's farm. Foster had only recently cleared his land, so he sowed corn among tree stumps in his field—farmers found it easier to remove the stumps from newly cleared fields if they left them to decay a while, then burned them. When the frosts hit, farmer Foster began burning his stumps a little sooner than originally planned. When stumps burn in the ground, the process is a long and smoky one as the roots burn deep into the soil. Foster's corn survived the cold summer in the warmth of the stump fires, and there was seed-corn in Fairfield for the following year.

Bankers from St. Albans offered five dollars a bushel, but Foster refused to sell his corn to them. Instead, the next spring, when people came to buy seed corn, he sold it to them for one dollar a bushel. Those who received the life-sustaining ears of corn called the place Egypt after the biblical story of Jacob, who sent his sons to Egypt to buy corn during the famine in Canaan.

manship and engineering savvy of their forefathers, have frequently tested the old bridges with prodigious loads. The Jaynes Bridge handled countless crossings until finally a truckload of gravel was too much. The nearby Lumber Mill Bridge swallowed a snow plow. The Montgomery Bridge strained and burst under a load of asphalt road-patch.

These happenings foretold the coming of the ubiquitous I-beam. Steel beams began to appear in the 1960s and 1970s as the old bridges were strengthened to support modern commercial traffic. The decks of the Church Street Bridge, the Montgomery Bridge, the Jaynes Bridge, and the Lumber Mill Bridge are all supported by steel beams. While the original trusses no longer support the roadways, these historic bridges remain well worth visiting.

Waterville

The Town of Waterville, originally chartered as Coits Gore in October 1788, was first represented at the legislature in 1829 by Luther Poland, forebear of Judge Luke Poland, the distinguished champion of the bridge at Cambridge Junction in 1885.

As the name implies, Waterville was a mill town—three or four large woolen factories and an equal number of other mills were located here. One of the woolens mills, described by contemporaries as "mammoth," was built and run by John Herrins of Ireland. It burned in the winter of 1853. Another woolens factory owned by Robert Herrins burned in December 1860.

Church Street Bridge

The view through the Church Street Bridge, up Church Street toward the village, frames the Waterville town hall. The setting in the middle of a small New England village makes the little bridge a favorite of photographers and artists.

The Church Street Bridge or Village Bridge features board and batten siding and rounded portals, differing from its more austere sisters to the north. The date of build and the name of the builder is unknown. Robert

Church Street Bridge

Town: Waterville
Date: c. 1877
Truss length: 60'
Builder: Unknown
Truss type: Queenpost

Waterville's village bridge stands over the North Branch at the bottom of Church Street. The setting captures the essence of life in a small New England town and invites viewers to stroll up the tree-lined avenue.

Hagerman, in his *Bridges of Lamoille County,* guesses the date to be sometime around 1877. This date is also used by the Vermont Division of Historic Sites in nominating the bridge for inclusion in the National Register of Historic Places. The Agency of Transportation lists the date as 1895.

A truck went through the deck in the winter of 1967. When the bridge was repaired in the following year, the roadway was made self supporting with the addition of four steel beams.

The sixty-foot queenpost truss span crosses the North Branch of the Lamoille River in the center of the Village of Waterville just to the west of Route 109.

Montgomery Bridge

The Montgomery Bridge's surroundings are park-like and especially attractive in fall. To the north, the river flows through ridges of bedrock. On the west bank upstream, a boulder reminiscent of an upended mill stone perches among the trees on the stream's edge.

THE LAMOILLE RIVER AND THE NORTH BRANCH

The Montgomery Bridge crosses the North Branch of the Lamoille River 1.2 miles north of the Village of Waterville, next to Route 109. The sixty-three-foot queenpost truss bridge is named for the Dallas Montgomery farm it serves. It is also known as the Lower Bridge in reference to the Upper, or Jaynes Bridge, upstream. This may be the bridge referred to by the *Cambridge Transcript* when it reported in November of 1887 that the Potter Bridge was completed.

The bridge was reconstructed with a self-supporting roadway in 1971 after a truck loaded with asphalt collapsed the deck. The truss is now required to support only itself.

Montgomery Bridge	
TOWN:	Waterville
DATE:	1887
TRUSS LENGTH:	63'
BUILDER:	Unknown
TRUSS TYPE:	Queenpost

The Montgomery Bridge stands in park-like surroundings that are especially attractive in the fall. Families come here in the summer to wade, swim, and picnic among the bedrock formations upstream of the bridge.

The Jaynes Bridge, sometimes known as the "Kissing Bridge," provides the only access to the farms and dwellings of Codding Hollow. The road it serves ends at the famous Long Trail.

Jaynes Bridge

The Jaynes Bridge is a classic example of a northern Vermont covered bridge. The functional lines of the bridge, rural setting, winding road, and background of mixed hardwoods and conifers provide something special to satisfy the viewer in any season.

A weathered sign on the west gable-end proclaims this to be the "Kissing Bridge," a claim that is probably true, since many of Vermont's covered bridges enjoy that reputation. It is named after a family who lived and worked nearby. Located in Codding Hollow, it has also been called the Codding Hollow Bridge, and it is also known as the Upper Bridge, as it is upstream of the Montgomery Bridge.

The bridge is located next to Route 109 about 1.7 miles north of the Village of Waterville. The fifty-seven-foot queenpost truss spans the North Branch of the Lamoille River, serving Codding Hollow Road and providing the only access to the farms and dwellings on the east bank of the river. The road continues another two miles to connect with the Long Trail as it threads its way through the foothills of the Green Mountains.

Jaynes Bridge

TOWN: Waterville
DATE: c. 1877
TRUSS LENGTH: 57'
BUILDER: Unknown
TRUSS TYPE: Queenpost

> **Covered Bridges and the Three R's**
>
> How, you may ask, can a knowledge of covered bridges, relics of the nineteenth century, help today's children learn to compete in tomorrow's world? The teachers and students of a fifth grade class at the Camels Hump Middle School in the Town of Richmond can tell you.
>
> Teachers Susan Girardin and Kerry Young designed an eight-week teaching unit titled "Bridges: Spanning Time." For four of the eight weeks, the focus is on covered bridges. The unit features visiting experts, a toothpick bridge contest, and a field trip to study covered bridges. All of the subjects a fifth grader is required to master are made interesting by being related to some aspect of a bridge.
>
> For the contest, the class is organized into toothpick bridge construction teams. Each group's task is to complete a toothpick bridge of their own design according to building codes and specifications. Weight is added to a bucket hung at the middle point of each bridge until the bridge collapses. The bridge that holds the most weight is the winner. The kids in the first group to be taught the unit learned truss construction too well—their bridges were tested with much more weight than had been planned, but some of them still just would not collapse.

The build date and the names of the builders are unknown, but the Division of Historic Sites estimates the build date to be circa 1877. The wooden roadway was made self supporting with four steel beams after a truckload of gravel broke through the original deck in 1960.

BELVIDERE

Belvidere was one of five grants made by the State of Vermont to John Kelly of New York by a charter issued November 1791. The town lies in the heart of the Green Mountains. Except for a few farms along the river valley, the terrain is suitable only for forest industries. The river, the North Branch of the Lamoille, was once named the Kelly River for the town's grantee.

In the 1800s there were several mills active in all seasons, and the manufacture of shingles, lathe, butter tubs, and sap buckets was carried on. It was said in those years that every other man in Belvidere was a cooper—a maker of barrels and tubs.

There are three hamlets in the town: Belvidere Center, Belvidere Corners, and Belvidere Junction, the last named for an expected railroad that never came.

Lumber Mill Bridge	
Town:	Belvidere
Date:	c. 1895
Truss length:	70'7"
Builder:	Lewis Robinson
Truss type:	Queenpost

Belvidere is home to two covered bridges: the Lumber Mill Bridge and the Morgan Bridge. Both cross the North Branch of the Lamoille River. To find them, drive north 3.7 miles on Route 109 from Church Street in Waterville. Cross a small cement bridge and turn left onto Mill Bridge Road. Drive 0.4 miles to the Lumber Mill Bridge. Cross the bridge and continue on, bearing right at the first fork. Drive about 0.8 miles to the Morgan Bridge. Cross the Morgan Bridge to return to Route 109.

Lumber Mill Bridge

The Lumber Mill Bridge, named for one of the mills that once lined the river here, is remarkable for its setting. There is a rocky ledge near its southwest corner with a cover of conifer trees. To the east, the river flows through large stone formations worn smooth by the water. The shoreline is easily accessed for viewing the bridge from below and the rock formations upstream. The foundations of the lumber mill can be seen along the river bank east of the north portal.

The Lumber Mill Bridge spans a rocky bend in the North Branch of the Lamoille River. The foundations of the mill the bridge was named for can still be seen on the north bank, downstream of the bridge.

Look for the several holes four to eight inches in diameter drilled in the bedrock near the bridge. One can only speculate on what purpose they once served.

The exact build date is undetermined but is probably in the middle 1890s—the Vermont Division of Historic Sites assumes the year to be 1895. Lewis Robinson built the seventy-one-foot span.

In 1971 a snowplow went through the bridge-deck. When repairs were made soon after, the bridge was reconstructed with an independent timber deck roadway supported by four steel beams. The rest of the structure is original, but the original trusses were repaired in 1995. A queenpost brace was replaced, as were large sections of both lower chords. Bridge restorer Jan Lewandoski and post-and-beam carpenter Paul Ide performed the work.

Morgan Bridge

A stand of pines mutes the sounds of the busy highway above the Morgan Bridge. The sixty-four-foot queenpost truss span was built by Lewis Robinson,

The Morgan Bridge is named for a family that owned the property at the north portal. A covered bridge inspection team found that the truss incorporates some unique features.

Morgan Bridge

Town: Belvidere
Date: 1887
Truss length: 63'8"
Builder: Leonard, Robinson, Tracy
Truss type: Queenpost

Charles Leonard, and Fred Tracy in 1887. It is named for a family that owned the property across the road at the north end.

An Agency of Transportation inspection report noted that the bridge has some unique design features. "The queenpost truss incorporates three small king-rod trusses within the queenpost truss to help support the floor loads. Also, queen-rods are positioned next to the queenposts. Two other short rods drop from near the bottom of the queenpost main-braces as well. Another design feature includes double six-by-eight tie beams at each queenpost allowing for two tenons and two pairs of knee braces."

The bridge also features a five-foot gable-end overhang at each end, unique among the queenpost bridges of the North Branch of the Lamoille.

Sources

Fairfax Bicentennial Committee. *Fairfax Vermont—Its Creation and Development, 1776 to 1976.* 1976.

Fairfield Bicentennial Committee. *Fairfield Vermont Reminiscences, 1763-1977.* 1977.

Hagerman, Robert L. *Covered Bridges of Lamoille County.* Essex Junction, VT: Essex Publishing Co., 1972.

Hemenway, Abby Maria. *The Vermont Historical Gazetteer, Vol. III.* Burlington, VT: A. M. Hemingway, 1877.

Historical Records Survey. Inventory of the Town, Village, and City Archives of Vermont, #8, Lamoille County, Volume I, Town of Belvidere. Montpelier, Vermont, 1940.

Lewandoski, Jan. *Wood Truss Highway Bridges in North America: Repair and Strengthening.* Proceedings of the Fifth International Conference on Structural Faults and Repair, pp. 217-223, Edinburgh, Scotland, 1993.

Nelson, Joe. "Fun and Lessons in Old Covered Bridges." *The Mountain Villager*, Vol.VII, No. 12, June 10, 1993.

Noble, Winona S. *The History of Cambridge Vermont.* Cambridge VT: Crescendo Club Library Assn., 1976.

Vermont Agency of Transportation Covered Bridge Study. Prepared for the State of Vermont Agency of Transportation by McFarland-Johnson, Inc., Binghamton, New York, 1995.

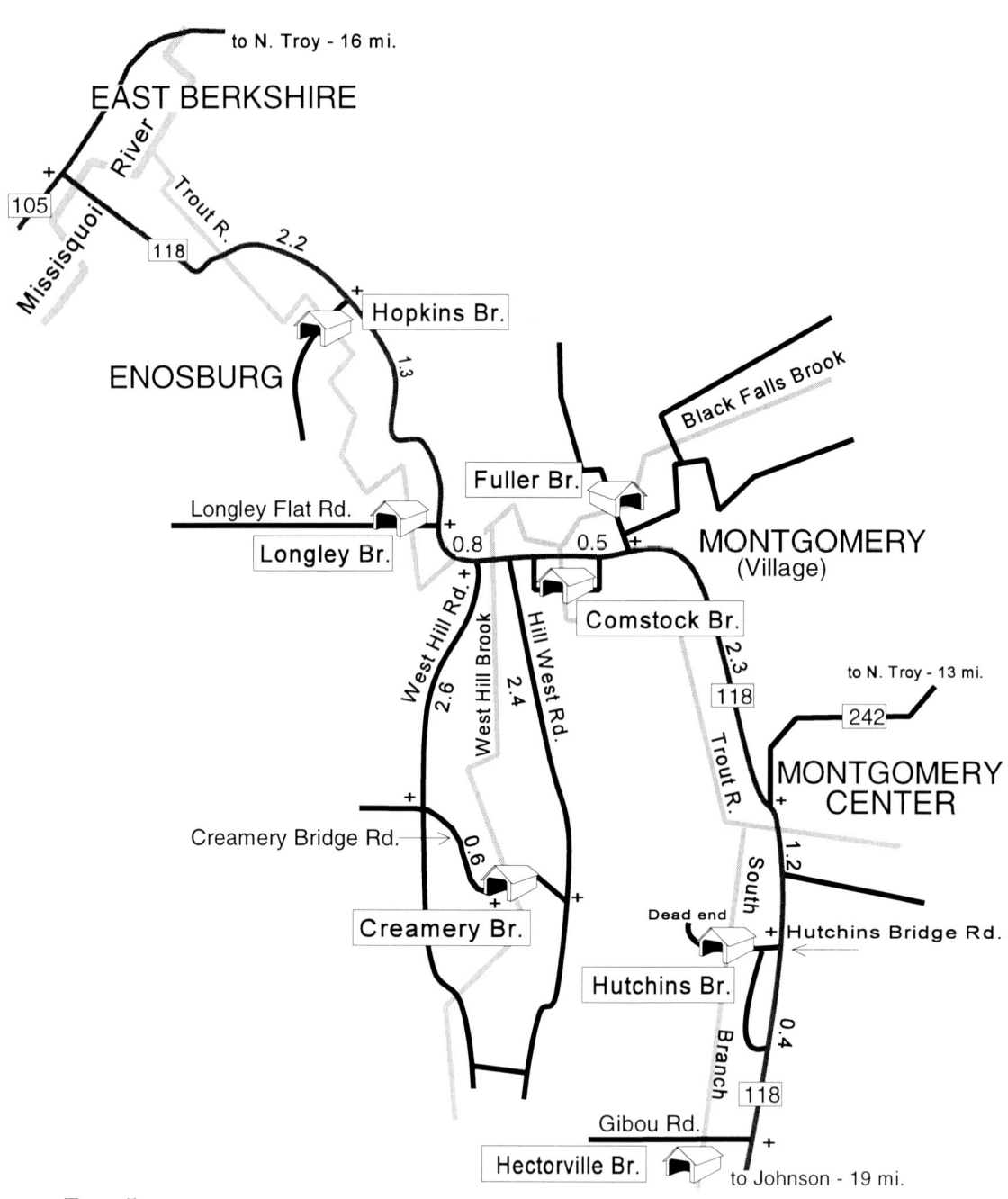

Tour 5
+ = *Distance marker*
Distances in miles (estimated)
Map not to scale
© *1996 Joseph C. Nelson*

The TOWN of MONTGOMERY

Tour 5

Montgomery, named for Revolutionary War hero Richard Montgomery of New York, was chartered in October 1789. General Montgomery had commanded American forces under Schuyler in the expedition against Canada. The Americans took Montreal but were defeated in front of Quebec City, where Montgomery was killed.

The Town of Montgomery is home to seven wooden bridges, all built by the brothers Savanna and Sheldon Jewett[1] over a period of about thirty years. In all, the Jewetts built nine plank-lattice bridges in the Montgomery area. They used timber from the family farm on West Hill and dressed the lumber in their own mill.

The northern edge of the Green Mountain chain rises around the Town of Montgomery on three sides. The valleys among the foothills are laced with small rivers and streams. Fed in the spring by melting mountain snows and in the summer by the frequent rains that make the Green Mountains green, the streams usually keep mill ponds filled for much of the year. With a wealth of water power and accessible softwood forests, Montgomery became a town of mills, turning spruce and hemlock into sap buckets and butter tubs, bobbins, and veneer.

Factories sprang up by the streams, sharing the waters. The peak growth years for the mills stretched from

---

[1] There were eight Jewett brothers: Braman, Giles, Oscar, Samuel, Alfred, William, Savanna, and Sheldon.

the 1860s into the 1890s, the years the Jewetts built their bridges. Each new mill created the need to cross the stream that powered it, so the boom years of mill construction led to a flurry of bridge building as well.

Montgomery Center

There are two settlements in the town, about two miles apart—Montgomery Village and Montgomery Center. Montgomery Center, the seat of the town government, was a mill town. The most famous enterprise may have been the Nelson and Hall Company veneer factory. Nelson and Hall operated the mill that made the veneer for the Victrola Company's cabinets. The mill-works used to stand across the Trout River from the Route 118 bridge near the junction of Route 242. A scattering of other mills operated along the South Branch of the Trout River. Jewett-built covered bridges stand near two of these old mill sites.

Hectorville Bridge

A stout wooden fence blocks the portals of the Hectorville Bridge. After serving Gibou Road since the turn of the century, it stands in retirement beside a concrete and steel replacement.

Gibou Road leaves the west side of Route 118 approximately six miles north of the junction of Route 109 at Belvidere Corners, or 1.6 miles south of the junction of Route 242 in downtown Montgomery Center.

Originally built in Montgomery Village in 1883, the bridge was moved to its present site in 1899 to serve the commercial bustle of a tub factory now long gone. There is easy access to the stream bed and a beautiful view of the bridge and the river. Upstream, the South Branch of the Trout River winds its way through a tumble of huge stones. Downstream, the river drops into a ravine in a spectacular waterfall.

At some point in time, a log was slung under the center of the span. Timbers were added to each truss in the form of an inverted V, and the log was suspended at its ends by steel cables from each timber apex. It was an effort to strengthen the span with a form of the kingpost truss. The result is the only plank-lattice/kingpost hybrid in Vermont.

Hectorville Bridge

Town: Montgomery
Date: 1883
Truss length: 52'9"
Builder: S. & S. Jewett
Truss type: Plank-lattice

The Hectorville Bridge spans a small gorge carved by the South Branch of the Trout River. With the lower chords rotted through, a jury-rigged kingpost keeps the old span from dropping into the ravine.

Hutchins Bridge

The Hutchins Bridge stands in a quiet valley out of view of the busy highway. A narrow unpaved road leads the traveler to the portal of the barn-red bridge. It is easy to imagine, in this isolated spot, that one has returned to the nineteenth century.

The Agency of Transportation Covered Bridge Study team inspected the bridge in 1994. The team found it to be in such poor condition that they recommended prompt attention to restore its capacity to safely support traffic.

The Tub Factory

It would be wrong to think of the Hutchins Bridge's history as idyllic and bucolic. This was a busy spot in 1883, as the moss-covered foundations just south of the west portal reveal. Here, Joseph Hutchins' five-lathe tub factory produced 2,000 butter tubs a day. The bridge resounded with the arrival of the mill workers at dawn, and again with their departure at dusk. Teams of horses clattered through, bringing logs of spruce, hemlock, and basswood, and other teams took the completed butter tubs away. The bustle stopped only when night fell. Over the years, as industry and society changed, the activity waned, until it finally stopped altogether when the factory shut its doors for good.

Hidden by foliage in its isolated hollow, Hutchins Bridge stands next to the ruins of the butter tub factory it once served.

The Agency asked the town to close the bridge until repairs are performed.

The bridge serves the dead-end Hutchins Bridge Road, leaving the west side of Route 118, 1.2 miles south of the junction with Route 242. Turn west here and you will find a crossroad. Drive straight through, downhill, and around a curve to the Hutchins Bridge.

Montgomery Village

White painted portals and splash-panels give the three bridges in and near Montgomery Village a clean and cared-for look. The Fuller Bridge, the Comstock Bridge, and the Longley Bridge are each named with wooden cutout letters placed high on the gable-ends, additional evidence of tender loving care. Except for the addition of distribution beams tie-bolted under the decks of all three bridges, the structure of the bridges remains as the builders left them. A fourth span, the Creamery Bridge,

---

Hutchins Bridge

Town: Montgomery
Date: 1883
Truss length: 77'
Builder: S. & S. Jewett
Truss type: Plank-lattice

is unpainted, but like all of Montgomery's covered bridges, it is marked with a signboard testifying to the Jewetts' craftsmanship and the year of construction.

Fuller Bridge

> **Fuller Bridge**
>
> TOWN: Montgomery
> DATE: 1890
> TRUSS LENGTH: 49'8"
> BUILDER: S. & S. Jewett
> TRUSS TYPE: Plank-lattice

The Fuller Bridge, or Black Falls Bridge, stands in the heart of Montgomery Village and is very much part of village life. In season, it can be found decorated with Christmas wreaths and lights.

The Fuller Bridge crosses Black Falls Brook in sight of Route 118. Black Falls Brook powered the J. E. Smith bobbin factory, and the Jewetts built the fifty-foot span in 1890 to replace an open bridge that had collapsed under the weight of a load of bobbins from the mill. The company built a mill town at the end of the present-day Black Falls Road, complete with a fifty-room dormitory and twenty-six mill houses, but no trace of it remains.

The bridge was restored in 1981, but it needed attention again soon thereafter. In early winter 1982 Montgomery's road crew knew there was a problem when a

The Fuller Bridge once served the J. E. Smith bobbin factory. The Jewett brothers built it to replace an open bridge that collapsed under a load of bobbins.

THE TOWN OF MONTGOMERY

> **Crystal Palaces**
>
> A longtime Montgomery resident recalled crossing the Fuller Bridge in the winter with his uncle to go to the dam at the bobbin works. There, where wooden penstocks brought water under tremendous pressure to the mill wheels, his uncle would drill several holes to create fountains up to forty feet high. The fountains of water froze and formed crystal palaces that people would come from near and far to see.

log truck passed the town garage with Christmas lights and pieces of roof rafter clinging to the log boom and dragging behind. There were pieces of roof scattered for half a mile, the road foreman said. "We had just put the lights up for the season." The trucker was obviously feeling very little pain when he drove through the bridge—he scarcely noticed when his boom hooked into the sway braces and dragged the roof off. The bridge served through the winter with the sides braced up with cables and was repaired the next year.

The Fuller Bridge stands next to Route 118, 2.3 miles north of the junction of Route 242 in Montgomery Center. Continue north on Route 118 about a half-mile to find the Comstock Bridge.

Comstock Bridge

The Comstock Bridge crosses the Trout River and provides alternate access to Route 118 for a few homes. It is located on Comstock Bridge Road, a piece of the old highway to Berkshire. The surroundings here are park-like through the efforts of local property owners.

The Comstock Bridge features a side port unique to the area—it exposes the lattice truss and lends the bridge a trim and jaunty look. Not installed for appearance, the port allows bridge users a glimpse around a sharp bend in the road. The abutments are of irregular stone laid without mortar. The west abutment stands on exposed bedrock.

The span was built in 1883 near the mill works of entrepreneur John Comstock, a miller, grain dealer, and manufacturer of carriages and sleighs.

> **Comstock Bridge**
>
> TOWN: Montgomery
> DATE: 1883
> TRUSS LENGTH: 68'10"
> BUILDER: S. & S. Jewett
> TRUSS TYPE: Plank-lattice

A side port in the Comstock Bridge exposes the plank lattice truss and gives the user a glimpse around a sharp bend in the road. Look up through the rafters to see the original wooden shingles under the metal roof.

Creamery Bridge

TOWN: Montgomery
DATE: 1883
TRUSS LENGTH: 58'8"
BUILDER: S. & S. Jewett
TRUSS TYPE: Plank-lattice

Creamery Bridge

The first glimpse of the Creamery Bridge is of the rooftop peeking through a sheltering screen of hemlock and black birch. The little bridge, spanning fifty-nine feet, crosses West Hill Brook high above a cascade of crystal clear water deep in a hollow on the south slope of West Hill.

In 1883, when the bridge was built, West Hill was a busy place. Besides the Jewett family farm and dimension lumber mill, there were forty-nine active farms. A creamery stood just east of the bridge, and there was a furniture factory in the lower West Hill Brook gorge.

Structural problems forced closure of the bridge in the summer of 1994. The covered bridge inspection team recommended interim structural rehabilitation to avoid collapse of the structure under its own weight and from snow loading.

The site is found 2.5 miles up an unpaved road not shown on most maps. An attempt to go there during Vermont's famous mud season would prove to be an adventure. There are two roads leaving the south side of Route 118 west of Montgomery Village, one at each end

of the new cement bridge. One is named West Hill Road and the other is Hill West Road—perhaps some Yankee humor is at work here. Take either road—both pass the bridge, and they ultimately meet in the valley below.

Longley Bridge

The Longley Bridge stands crisp and clean, with new deck, roof, and siding. The eighty-five-foot span was restored during the fall and winter of 1992 by Jan Lewandoski. Also called the [Samuel] Head Bridge, it crosses the Trout River north of the village, serving Longley Flat Road, giving access to the Enosburg Town Forest and East Enosburg.

In 1979 the stonework was capped and faced with concrete, and additional support was given to the ends of the bridge stringers by means of three steel beams cantilevered from the base of the west abutment.

A waterfall and pool hidden beneath the Creamery Bridge lure local residents on hot summer days. The ruins of the creamery stand at the north portal, and the Jewett family farm and lumber mill were also located near here.

Longley Bridge

TOWN: Montgomery
DATE: 1863
TRUSS LENGTH: 84'7"
BUILDER: S. & S. Jewett
TRUSS TYPE: Plank-lattice

The Town of Montgomery recently contracted with Jan Lewandoski to restore the Longley Bridge. The youngsters who come here for fishing and swimming quickly made adjustments to the new siding.

The Longley Bridge lies west of Route 118, 0.8 miles north of the intersection of West Hill Road.

Enosburg and the Hopkins Bridge

The Enosburg Township was granted March 12, 1780, and chartered May 15, 1780 to Roger Enos and fifty-nine associates.

The Hopkins Bridge stands just over the Enosburg-Montgomery town line off Route 118, 1.3 miles north of the Longley Bridge. The portal is unpainted and there is no name plate, but it is distinguished by a carefully lettered formula sign that reads: "Slow Autos to 10 Miles an Hour Horses to a Walk Per Order Selectmen."

Enosburgh Post Office, now Enosburg Center, and East Enosburg with East Berkshire and Richford. If there was any other reason for the bridge's existence, the evidence is gone. Any dams or foundations that might have been here have been erased from the rich bottom land lying along the twisting and turning course of the Trout River.

Today the bridge serves a farm on a dead-end road. Some local residents also make use of it as a message center. It was closed and bypassed with a temporary bridge when, in 1993, the Agency of Transportation Covered Bridge inspection team found it to be "severely over stressed." The report recommended general structural rehabilitation.

"Ironically, or coincidentally," the report stated, "rehabilitation to provide sufficient capacity for self weight and snow loading will also provide sufficient capacity to reopen the bridge for vehicular traffic. Failure to rehabilitate risks the collapse of the [structure]." Before anything can be done, of course, the town must first find the necessary funds.

SOURCES

> **Hopkins Bridge**
>
> TOWN: Enosburg
> DATE: 1875
> TRUSS LENGTH: 90'9"
> BUILDER: S. & S. Jewett
> TRUSS TYPE: Plank-lattice

The Hopkins Bridge stands near a bend in the sandy Trout River. The span once served the old road to Berkshire, now abandoned. At ninety-one feet, it is the longest of the Jewett bridges.

Sources

Allen, Richard Sanders. *Covered Bridges of the Northeast.* Brattleboro, VT: Stephen Greene Press, 1957.

Branthoover, W. R. and Taylor, Sara. *Montgomery Vermont: The History of a Town.* Burlington, VT: The Montgomery Historical Society, 1976.

Congdon, Herbert Wheaton. *The Covered Bridge.* Middlebury, VT: Vermont Books, 1973

Geraw, Janice Fleury. *Enosburg, Vermont.* Enosburg Falls, VT: Enosburg Historical Society, Inc., 1985.

Hemenway, Abby Maria. *The Vermont Historical Gazetteer, Vol. II.* Burlington, VT: A. M. Hemenway, 1868.

Vermont Agency of Transportation Covered Bridge Study. Prepared for the State of Vermont Agency of Transportation by McFarland-Johnson, Inc., Binghamton, New York, 1995.

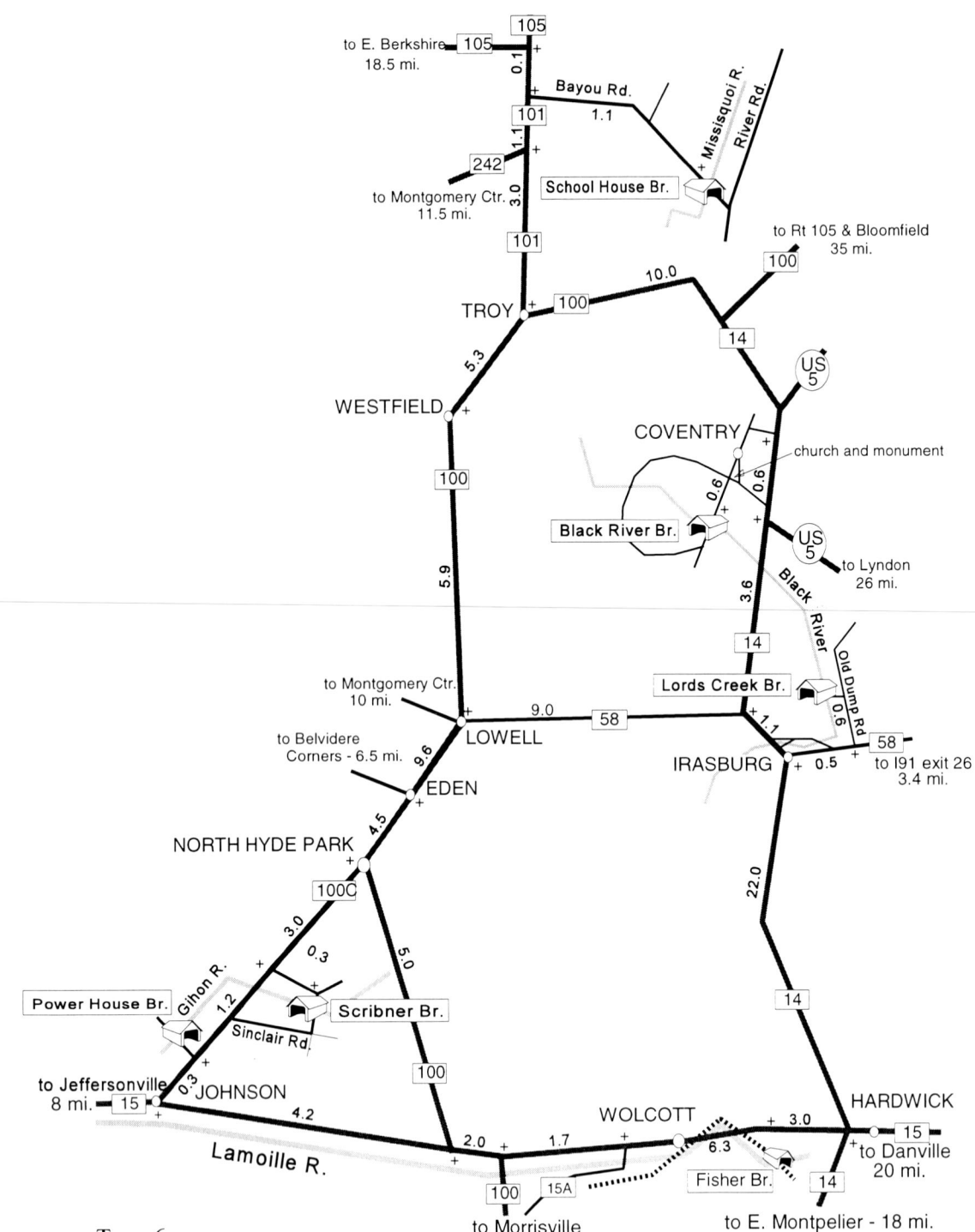

Tour 6
+ = Distance marker
Distances in miles (estimated)
Map not to scale
© 1996 Joseph C. Nelson

ROUTE 100 in NORTHERN VERMONT

Tour 6

There are six covered bridges of five different construction types scattered along the roughly thirty miles of Route 100 between North Troy and Route 15. The countryside varies from mountain to farmland. The bridge-site searches lead off the beaten track deep into Vermont's Northeast Kingdom and through some of the state's most picturesque villages.

Because the Route 100 tour lies in the center of northern Vermont surrounded by five other tours, it can be entered from any one of these or from Interstate 91. For the purpose of organization of the text, entry is assumed to be made from Montgomery and Route 242. The bridge descriptions are laid out from North Troy south to Wolcott. With this in mind, bridge viewers are invited to use the tour map for this chapter to plan their own itineraries.

TROY

Troy was organized in March 1802 as Missisco, after the river Missiscoi, as it was spelled then. The Indian name for the Missisquoi River is Azzastaquake, describing a river that turns back on itself. The name Missisquoi, a corruption of Masseepsque meaning "the place of arrow flints," actually applied only to the Missisquoi Bay area, not the river. Settlement began in 1796, when several men from Peacham, Vermont, came to explore the area. The first settler arrived in 1797.

The town name was changed to Troy, not in honor of the ancient city written about by the Greek poet, Homer, but after Troy, New York. It seems that Vermonters who were in debt to New York creditors were at risk of imprisonment when they crossed the state line. Some Vermonters from Missisco, arrested for debt in New York, were released on bail put up by Troy, New York, resident Benjamin Smith. In gratitude, they named their town after his.

School House Bridge

Barn red, the ninety-two-foot School House Bridge stands at a bend in the Missisquoi River, next to a ford and a sandbar. At low water, a viewer can walk out on the sand to within a hundred feet of the span.

The bridge is of special interest because of the three pairs of flying buttresses and because of the low roof line. The height of the plank-lattice trusses on this bridge are only nine feet eight inches, one to two feet shorter than usual, leaving insufficient space for getting hay wagons through if the conventional interior bracing system had been used. The external buttresses provide the required lateral bracing to resist wind forces and to keep the bridge trusses square and straight.

The trusses are of comparatively light construction—the lattice planks measure two inches by nine inches versus the usual three inches by ten or eleven inches, and the web crossings are pinned by one treenail where normally two are used. Further, three sets of chords strengthen the span rather than the usual four, as the upper secondary chords are missing. The chords also consist of only one plank on each side of the webbing instead of the usual two. On the positive side, the chord planks are twice the thickness of those used in other plank lattice bridges, measuring at least six inches. Where other lattice-bridge chords use two planks on each side of the web, staggered and pinned to gain length, the chords in the School House Bridge are joined end to end using a peculiar "E" shaped splice block secured with bolts. Also on the positive side, the web spacing is on twenty-nine-inch centers instead of the customary thirty-six-inch center.

School House Bridge

TOWN: Troy
DATE: 1910
TRUSS LENGTH: 92'5"
BUILDER: Unknown
TRUSS TYPE: Plank-lattice

The School House Bridge is one of the few bridges in Vermont to use external lateral bracing. It is also the only plank-lattice truss bridge in the state to use only one treenail where the lattice planks cross.

Using just one treenail at the web intersections "may be a weakness," said an Agency of Transportation inspection report, "since some compression lattice members are buckling away from the tension members with gaps developing between the two."

The Vermont Division of Historic Sites used a bridge build date of 1910 in the application for inclusion of the bridge on the National Register of Historic Places. The name of the builder is unknown.

According to the research of Richard Sanders Allen, there were formerly three similar single-pin buttressed bridges in the towns of Troy and Westfield. Phil Ziegler, in his *Sentinels of Time: Vermont's Covered Bridges,* tells us that the School House Bridge collapsed in 1958 and was later rebuilt. The bridge also survived a fire, date unknown. The structure shows signs of the sandblasting done to remove the char.

School House Bridge is found on Bayou Road, 1.1 miles from the junction with Route 101. Bayou Road leaves the east side of Route 101, 1.1 miles north of the junction of routes 242 and 101, just south of where Route 105 joins Route 101.

To see a spectacular waterfall, pass through the bridge and turn left onto River Road. Drive 1.2 miles to Big Falls, where the gorge and cataract can be viewed from atop a rock pinnacle. To continue the tour to Coventry, retrace Bayou Road to Route 101 and go south 4.1 miles to Route 100 at Troy Village. Take Route 100 ten miles east to Coventry, passing the junction with Route 105 and taking Route 14.

COVENTRY

Chartered to Major Elias Buel and fifty-nine others in 1780 by the State of Vermont, Coventry was named after Buel's birthplace in Connecticut. At the first town meeting, held in 1803, it was voted that each inhabitant should work on the roads four days in June and two days in September. This practice was not at all unusual in the early years of the United States. In cash-poor communi-

The Black River Bridge, also known as the Coventry Bridge, is one of only three spans in Vermont using the intricate Paddleford truss and the only one serving daily traffic. The selectmen must frequently replace the planks in the gable-ends, as passing trucks just as frequently punch them out.

82 VERMONT'S COVERED BRIDGES

ties, taxes and other debts were paid "in kind," meaning produce and labor were accepted as the medium of exchange.

In the fall of 1805 the first public roads were laid out. Until that time the roads were paths cut through the woods no wider than what was needed for a single team, and not always that wide. The roadway was cleared of trees, but stumps, stones, and mud holes remained for the traveler to avoid as best he could.

The town grew along the Black River. By 1840 there were two sawmills, a gristmill, a clothiery works, a tannery, and a starch factory. Things were going so well that around the year 1840 there was a movement to make Coventry the shire town of Orleans County. The name of the town was actually changed to Orleans in 1841. The effort failed, and the name was changed back to Coventry in 1843.

Black River Bridge

The Black River Bridge, built in 1881, serves Coventry Road where the road follows the stream. Also known as the Coventry Bridge or Lower Bridge, it spans the Black River at a bend near a swampy shore, just over the Irasburg town line. Beavers work only a few feet from passers-by. In the springtime, ducks land in the wet fields off to the north.

The eighty-six-foot bridge is clean and well kept. It is one of three Paddleford-truss spans surviving in Vermont, and it is the only one supporting regular daily traffic. This bridge and the Lords Creek Bridge near the Village of Irasburg were built by John D. Colton of Irasburg. A third Paddleford structure, built in 1869 by unknown craftsmen, is located in Lyndonville.

The bridge stands high on original stone abutments now cased and capped in concrete. The interior is bright because of full-length venting. The siding on the south side stops short twenty-two inches from the eaves, while the siding on the north side leaves an open port eight feet high the length of the truss. The deck is not reinforced, and the floor system remains as the builder designed it. The town periodically renews the nicely rounded and scrolled portals, but to little avail—trucks

---

Black River Bridge

TOWN: Irasburg
DATE: 1881
TRUSS LENGTH: 85'10"
BUILDER: John D. Colton
TRUSS TYPE: Paddleford

passing through quickly demolish the work. The Agency of Transportation recommends that the bridge be closed to all trucks, bypassed, or rehabilitated to take all traffic.

To find the Black River Bridge, leave Route 14/Route 5 on the west-bound local road just south of where the two highways separate. Follow the local road southward to a large white frame church fronted by a Civil War monument. Drive to the right of the church and you will come to the bridge. Cross the bridge and continue on, taking the right-hand forks as you come to them. You will come around to the white frame church again by way of Heermansmith Farm Road.

To go on to Irasburg and the Lords Creek Bridge, leave Coventry on Route 14 south 4.2 miles to Route 58. Turn left on Route 58 and drive 1.6 miles through Irasburg to Old Dump Road. Drive north 0.6 miles to the bridge.

Irasburg

The Town of Irasburg was granted to Ira Allen by the Vermont General Assembly in February 1781. This Ira Allen was the nephew of Ethan and the son of Ira, the famous land speculator. A certain number of proprietors were needed to form a new township—probably sixty-two men in all. According to historian E. P. Colon, when the Allens wanted a new township granted, they collected a few genuine names towards the required number, then created the rest by inventing people from distant places. The Allens paid the state the first grantee dues and afterward "bought up" the claims of the fictional people.

Ira Allen resided in Irasburg from 1814 until his death in 1866.

Lords Creek Bridge

The Lords Creek Bridge in Irasburg is John Colton's other surviving Paddleford truss bridge. Built in 1881, it spanned Lords Creek until it was replaced in the late 1950s. The old bridge was acquired by the LaBonds and moved to the family farm, where it crosses the Black River. Still known as the Lords Creek Bridge, it provides access to the farmer's fields. There is a cattle gate stapled to the west portal.

Lords Creek Bridge

Town: Irasburg
Date: 1881
Truss length: 47'9"
Builder: John D. Colton
Truss type: Paddleford

Unfortunately, the forty-eight-foot bridge is in very poor condition. The siding and gable end sheathing is stripped off, and some of the bracing on the south side is pulled loose. In this open-sided condition, the structure is ideal for studying the mechanics of Peter Paddleford's truss.

Continue the tour by returning to the junction of Routes 58 and 14. Take Route 58 east nine miles to Lowell. At Lowell drive south 14.2 miles on Route 100 to North Hyde Park and the Route 100C junction. The Scribner Bridge is found on an unpaved road leaving Route 100C to the southeast, three miles from the junction with Route 100.

When the covered bridge crossing Lords Creek was replaced, it was moved over the Black River to serve the LaBond family farm. One of Vermont's last Paddleford truss bridges, it stands today the picture of utter neglect.

Johnson

Land for what was to become Johnson was first granted to a man named Brown. He had planned to name the town Brownsville, but he and his family were captured by Indians in 1780 and taken to Canada. When the charter fees were not paid, another grant was awarded to Samuel William Johnson in 1782. When the town was chartered ten years later, Johnson's name was chosen for the town.

The village grew up at the confluence of the Lamoille and Gihon rivers and, possessing good water power, soon grew into a thriving mill town. With the water power era long past, Johnson continues to be famous for the Johnson Woolen Mills.

Scribner Bridge

The Scribner Bridge is a pretty little bridge with pleasing proportions in a quiet rural setting north of Johnson Village. It stands in front of a cluster of farm buildings at the edge of wide fields cut by the Gihon River. Streamside trees mark the flow of the river in the distance.

The Scribner Bridge is named for a family that once worked the adjacent farm. It is also sometimes called the Mudget Bridge. There is evidence that the bridge is used to cross cattle to and from the fields.

The bridge is unique for its truss, the only one of its kind in Vermont. A modified queenpost arrangement without internal bracing, it employs iron rods instead of wooden posts to support the bottom chords and floor beams. The timber dimensions are comparatively massive for the size of the crossing—the horizontal mem-

> **Scribner Bridge**
>
> TOWN: Johnson
> DATE: 1919
> TRUSS LENGTH: 44'6"
> BUILDER: Unknown
> TRUSS TYPE: Queenpost (modified)

> **The Johnson Woolen Mills**
>
> The Johnson works are noted for their hunting clothing, particularly the familiar red and black plaid outfits that keep the wearer warm even when soaking wet. The mills have been in continuous operation since 1816.
>
> The original deed provided water privileges from the mill dam sufficient to run a fulling mill, to nap and shear, to cut die wood, and to dress cloth. Boosters advertised the mill as "Built on the Gihon that runneth westward from Eden," and, because the enterprise also owned a gristmill, they boasted that they "fed the hungry at one end of the dam, and clothed the naked at the other!"

The Scribner Bridge, simple and clean, stands in the river bottom lands of the Gihon. Probably built as an open bridge, it employs a one-third high queenpost-like truss using iron rods to support the stringers.

bers are ten by eleven inches, the diagonal members, ten by ten. The truss is only one-third of the height of the sidewalls. Each horizontal member supports three full-height posts, which in turn support the plates under the bridge roof. It is believed that the forty-four foot span began life as an open bridge. The Agency of Transportation Covered Bridge Report cites the year of construction as approximately 1919.

Reconstructed in 1960, the structure consists of an independent timber deck roadway supported by four steel beams. The original stone abutments were replaced with concrete.

Drive through the bridge and continue south on the unpaved road to the first intersection. Turn right to return to Route 100C. During mud season, maple syrup time, the viewer will find a sugar house in full smoke and steam. Turn left onto Route 100C and proceed one mile to the Power House Bridge.

Power House Bridge

The Power House Bridge was originally known as the School Street Bridge. It was built in 1870 to extend School Street across the Gihon River, but, when the village constructed an electric power plant upstream in the 1890s, the popular name changed to the Power House Bridge. The seventy-three-foot structure is a queenpost truss with massive timbers.

The bridge was reconstructed in 1960 and again in 1993 because the bridge was developing a decided sag. The truss was renovated, with much of the original timber replaced. Unfortunately, the bridge continued to sag, and it was closed again in 1995 for further work. The

> **Power House Bridge**
>
> TOWN: Johnson
> DATE: 1870
> TRUSS LENGTH: 60'10"
> BUILDER: Unknown
> TRUSS TYPE: Queenpost

The Power House Bridge's queenpost truss employs timber with dimensions generous for a bridge of this size. Conduits are carried across the Gihon River under the span.

bridge had been in daily use, the traffic flow heavy at all hours.

Continue south on Route 100C 0.3 miles to Route 15. Turn left, proceeding east 14.2 miles through the Village of Wolcott to the Fisher Bridge.

WOLCOTT

Chartered in 1781 by the Vermont Legislature, Wolcott was named for Major General Oliver Wolcott, one of the proprietors. The first town meeting was held in March 1791. "They elected all of their best men to office," wrote Henry Wiley for the *Vermont Historical Gazetteer*. "For all of the citizens in town were in office." The next recorded meeting took place in 1794, when there were but four voters in town. As a result, Thomas Taylor was elected town clerk, first selectman, and constable.

Fisher Bridge

The Fisher Bridge provides an example of government, business, and community working together to preserve private property in the interest of history. A marker put up by the Vermont Board of Historic Sites describes the bridge and why it is still with us:

> Fisher Bridge, Wolcott, Vermont. This bridge, spanning the Lamoille River on the St. Johnsbury & Lamoille County R.R., is the last railroad covered bridge still in regular use in Vermont and one of a very few left in the U.S. Built in 1908, it is the only one remaining with a full-length cupola, which provided a smoke escape. In 1968, the bridge was scheduled for destruction to make way for a new steel span. It was saved by placing heavy steel beams underneath. This preservation was achieved with State funds and with generous private donations raised by the Lamoille County Development Council.

The 103-foot bridge was built by the Pratt Construction Company, founded by Thomas W. Pratt, the inventor of the truss used in the Lincoln Bridge in Woodstock. The designers adapted the Town-lattice truss for the rail-

The Fisher Bridge is the last covered railroad bridge in use in Vermont. The large timber dimensions and doubled lattice of the Town-Pratt truss make it very sturdy—in fact, a bridge restorer called it probably the strongest wooden bridge in the state.

road and dubbed it the Town-Pratt truss. The adaptation beefed up the lumber dimensions and doubled the web on each side.

The preservation work involved adapting the abutments and adding a mid-span pier consisting of steel pilings driven into the river-bed. The original Town-Pratt truss no longer has a load-bearing function. The State has provided the site with a small park.

According to Robert L. Hagerman, in his *The Covered Bridges of Lamoille County,* the bridge serves a ninety-six-mile single-track line between St. Johnsbury and Swanton. The name, Hagerman relates, "is derived from Christopher Fisher, whose farm bordered the tracks and river crossing at the time the bridge was built."

At the time of this writing, the bridge is functional, but the railroad has ceased operations. The bridge is used occasionally during railroad tours conducted for railroad buffs and leaf peepers.

Fisher Bridge

Town: Wolcott
Date: 1908
Truss length: 103'
Builder: Pratt Construction
Truss type: Town-Pratt lattice

Sources

Allen, Richard Sanders. *Covered Bridges of the Northeast.* Brattleboro, VT: Stephen Greene Press, 1957

Butterfield, Anne Huckins. *Memories of the Early Days of Troy, Vermont.* 1977

Hagerman, Robert L. *The Covered Bridges of Lamoille County.* Essex Junction, VT: Essex Publishing Co., 1972

Hemenway, Abby Maria. *The Vermont Historical Gazetteer, Vol. II.* Burlington, VT: A. M. Hemenway, 1871.

Hemenway, Abby Maria. *The Vermont Historical Gazetteer, Vol. III.* Burlington, VT: A. M. Hemenway, 1877.

Historical Records Survey. Inventory of the Town, Village, and City Archives of Vermont, #10. Orleans County, Volume V, Town of Coventry. Montpelier, Vermont, 1940.

Vermont Agency of Transportation Covered Bridge Study. Prepared for the State of Vermont Agency of Transportation by McFarland-Johnson, Inc., Binghamton, New York, 1995.

Ziegler, Phil. *Sentinels of Time: Vermont's Covered Bridges.* Camden, ME: Down East Books, 1983

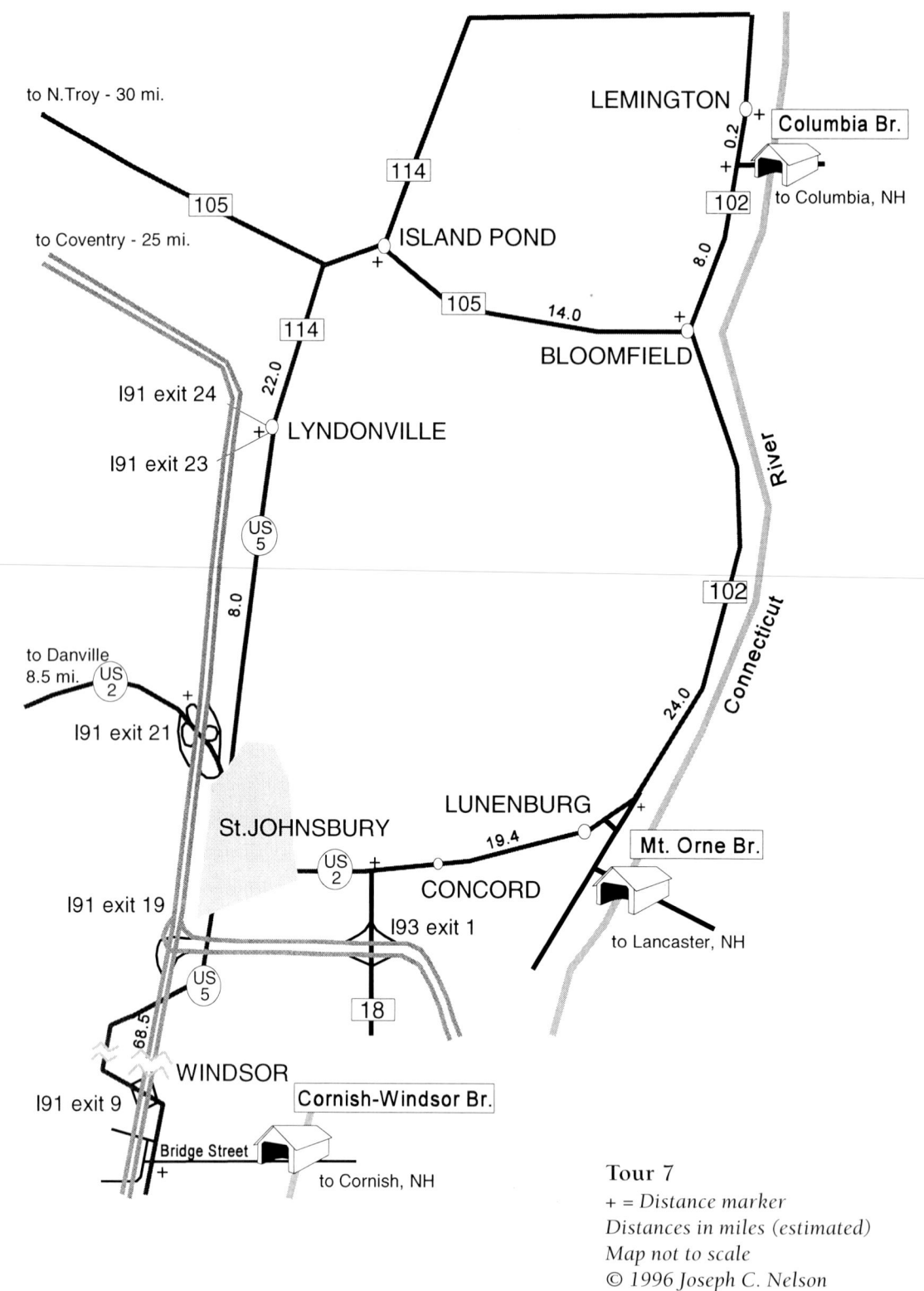

CROSSING the CONNECTICUT

Tour 7

The Vermont-New Hampshire state line follows the west bank of the Connecticut River. The boundary was established by King George III in 1764 as New York's eastern border and, despite the Revolutionary War, the bickering between the colonies, and Vermont's emergence as a state, the border remains right where the king put it more than two centuries ago. As a result, the western abutments of the Connecticut River bridges sit "just barely" in Vermont, and the only way to cross the Connecticut is on a New Hampshire bridge.

According to Richard Sanders Allen in his *Covered Bridges of the Northeast:* "At one time there were 35 covered highway and railroad bridges across the Connecticut between its source in the lakes of Pittsburg, New Hampshire, and the Massachusetts line at Hinsdale. As of June 1, 1957, there were five covered bridges spanning the Connecticut River in Vermont."

Today, there are but three covered bridges connecting New Hampshire and Vermont over the Connecticut River—they can be found at Lemington, Lunenburg, and Windsor. There would be four, but a storm that passed through Newbury in 1979 took down the Bedell Bridge. A resident recalled that a freak wind destroyed the bridge, then took off the roofs of the barns on a neighboring farm. The destruction occurred shortly after the State of New Hampshire had spent $250,000 on restoration. Visitors to the Bedell site will find only the old toll house—now a private residence—the bridge abutments, and a monument.

The Connecticut River bridges are a long distance apart. The northernmost bridge is within thirteen miles of the Canadian border, the southernmost stands more than one hundred miles south in Windsor. A visit to the Windsor Bridge could be delayed until the traveler is in the neighborhood.

Lemington

Lemington Town was granted by New Hampshire Governor Benning Wentworth to Samuel Averill and sixty-three others in 1762. Industry apparently had a slow start. According to the *Vermont Historical Gazetteer,* a gristmill was not built there until 1810. Until then, the settlers had to carry their grain twenty-five miles to Guildhall to be ground. Dependent primarily on agriculture and forest-related industry, the area remains rural. The 1990 census found only 102 people living near the Vermont end of the Columbia Bridge.

Columbia Bridge

The northernmost covered bridge crossing the Connecticut River from New Hampshire to Vermont is the 146-foot Columbia Bridge, named for Columbia, New Hampshire. It stands just a few feet from the east side of Vermont Route 102.

This is the third bridge on this site, built in 1912 by Charles Babbitt using the Howe truss. The first bridge was lost to a storm, and the second was set afire by a spark from a railroad locomotive. One might wonder why the State of New Hampshire would build bridges in this out-of-the-way place—there never has been much population here, and there are no connections to commercial centers. The incentive was the wealth of Vermont's northern forests. New Hampshire's mills gained access to Vermont timber, and jobs were created in the mountains and at the mills. The economies of the communities on both ends of the bridge have been linked from early times.

The Columbia's Howe truss has a clean, modern appearance, and represents the last word in wooden bridge

---

Columbia Bridge

TOWN: Lemington
DATE: 1912
TRUSS LENGTH: 146'
BUILDER: Charles Babbitt
TRUSS TYPE: Howe

The single-lane, single-span Columbia Bridge connects Columbia, New Hampshire, with Lemington, Vermont, where the mighty Connecticut River is still a small stream. This highway bridge uses the Howe truss, which is usually found in railroad bridges.

technology. The interior is well lighted by an open port in the north side. A constant flow of light truck and automobile traffic moves in both directions across the single-lane bridge, which has a six-ton load limit.

Lunenburg

Lunenburg was first settled circa 1768. Townspeople found Neal's Brook, Catsbow Brook, and Mink Brook to be good mill streams, and the alluvial valley offered superb farm land. Many mills were located here, giving rise to a forest products industry in addition to the agricultural activity.

Mount Orne Bridge

The 267-foot Mount Orne Bridge was built in 1911 by Charles Babbitt using the Howe truss. Mr. Babbitt built the Columbia Bridge at Lemington the following year using the same truss. The longer Mount Orne Bridge uses a mid-stream pier.

The Mount Orne Bridge, or Lunenburg-to-Lancaster Bridge, is located just south of a bend in the river. The

The Mount Orne Bridge stands in one of the Connecticut River Valley's most beautiful areas. Old photographs show a toll house standing against the south side of the Vermont portal.

view to the north may be among the most beautiful along the Connecticut River Valley. The tree-lined stream wends through rich bottom-lands among low rolling hills, and the mountains of New Hampshire loom in the distance.

The bridge is found on East Concord Road, about one mile south of where the road leaves Route 2. The junction is approximately twenty miles east of the I-93 clover or 1.5 miles east of Lunenburg where Route 2 bends north.

Windsor

Windsor is called "the birthplace of Vermont," for it was here that the constitutional convention assembled to form the state. It began with a charter issued on July 6, 1761, under the seal of the Province of New Hampshire, signed by Governor Wentworth. The town was named for John Stewart, Earl of Windsor.

On July 8, 1777, the constitution of the new state was signed by delegates trapped inside by a violent thunderstorm. The name New Connecticut was considered, but the proponents of the name Vermont carried the day.

Mount Orne Bridge

TOWN: Lunenburg
DATE: 1911
TRUSS LENGTH: 266'11"
BUILDER: Charles Babbitt
TRUSS TYPE: Howe

Cornish-Windsor Bridge

A historic marker stands on the New Hampshire side of the Cornish-Windsor Bridge. It says:

> Cornish-Windsor Bridge. Built in 1866 at a cost of $9,000, this is the longest wooden bridge in the United States and the longest two-span covered bridge in the world. The fourth bridge at this site, the 460-foot structure was built by Bela J. Fletcher (1811-1877) of Claremont and James H. Tasker (1826-1903) of Cornish, using a lattice truss patented by architect Ithiel Town in 1820 and 1835. Built as a toll bridge by a private corporation, the span was purchased by the State of New Hampshire in 1936 and made toll-free in 1943.

Cornish-Windsor Bridge

TOWN: Windsor
DATE: 1866
TRUSS LENGTH: 460'
BUILDER: James Tasker, Bela Fletcher
TRUSS TYPE: Timber-lattice

The first bridge was built in 1796 and lost to a freshet in 1824. The second bridge disappeared downstream in the spring of 1849. The third was lost to the flood of March 1866.

The new bridge structure was laid out in a meadow near Bridge Street in Windsor before it was assembled over the river. The builders used Ithiel Town's plan, but instead of planks, they used squared timbers in the lattice.

Within thirty years of construction the bridge displayed several inches of negative camber and disturbing vibrations under live loads. Several attempts to repair the problems did not succeed, and the bridge was closed for rehabilitation in 1987 by the State of New Hampshire, owner of 96 percent of the bridge—only 4 percent of the structure is located in Vermont.

The repairs needed to maintain the bridge's historic form and structural system, but the importance of the bridge to local transportation meant that they also needed to meet modern engineering standards. The structure's history of problems, engineering analysis, and experienced observation of the bridge indicated that it was overstressed by its own dead load and that merely repairing damaged members and restoring the bridge to as-built condition would not suffice. It was decided to use glu-laminated timbers, which have considerably higher de-

The present Cornish-Windsor Bridge is the fourth to stand at this crossing. When the bridge was closed in 1987 for rehabilitation, its importance was proven—places once conveniently close were now miles out of the way. For the two years it was closed, residents had to drive many more miles to conduct their business. The bridge has been a vital link between the two communities since 1866.

sign values in bonding and tension, for the chords. Glulaminated bolster beams were also added to support the truss under unusually heavy loads. The span was returned to service in the winter of 1989.

   The reconstruction became controversial among bridge historians when the contractors opted to use modern materials, such as glulaminated timber. The critics accused the engineers of contempt for the methods of the old bridge builders, while the modern designers pointed out that, according the basic principles of mechanical engineering, the old bridges should have collapsed. They had not done so, obviously, but it is understandable that the modern engineers would rather err on the side of caution.

   On the Vermont side, the Cornish-Windsor Bridge can be reached by turning east on Bridge Street in Windsor from Route 5. Travelers on Route I-91 should use Exit 9 and drive four miles south on U.S. Route 5 to Bridge Street.

The bridge supports two-way traffic. It must handle a lot of it during the rush hours in the morning and again in the afternoon, when the traffic flows almost unbroken in both directions. To see more of the bridge than a blur, avoid the rush hours.

On the Vermont side the street is narrow with still narrower sidewalks. Photographers should note that during the business day there is no safe place to stand to take pictures—the approach is so narrow that the only shot that can be framed is a straight-on view of the Vermont portal. There is no on-street parking, and the nearest parking place is a lot in front of an American Legion hall, half a city block away. The best place from which to view the bridge is from a small parking area on the New Hampshire side.

Sources

Allen, Richard Sanders. *Covered Bridges of the Northeast.* Brattleboro, VT: Stephen Greene Press, 1957.

Allen, Richard Sanders. *Rare Old Covered Bridges of Windsor County.* Brattleboro, VT: Stephen Greene Press, 1962.

Congdon, Herbert Wheaton. *The Covered Bridge.* Middlebury, VT: Vermont Books, 1973.

Hemenway, Abby Maria *The Vermont Historical Gazetteer, Volume I.* Burlington, VT: A. M. Hemenway, 1868.

Lewandoski, Jan. *Wood Truss Highway Bridges in North America: Repair and Strengthening.* Proceedings of the Fifth International Conference on Structural Faults and Repair, pp. 217-223, Edinburgh, Scotland, 1993.

Wardner, Henry Steele. *The Birthplace of Vermont: A History of Windsor to 1781.* New York: Charles Scribner's Sons, 1927. (Privately printed)

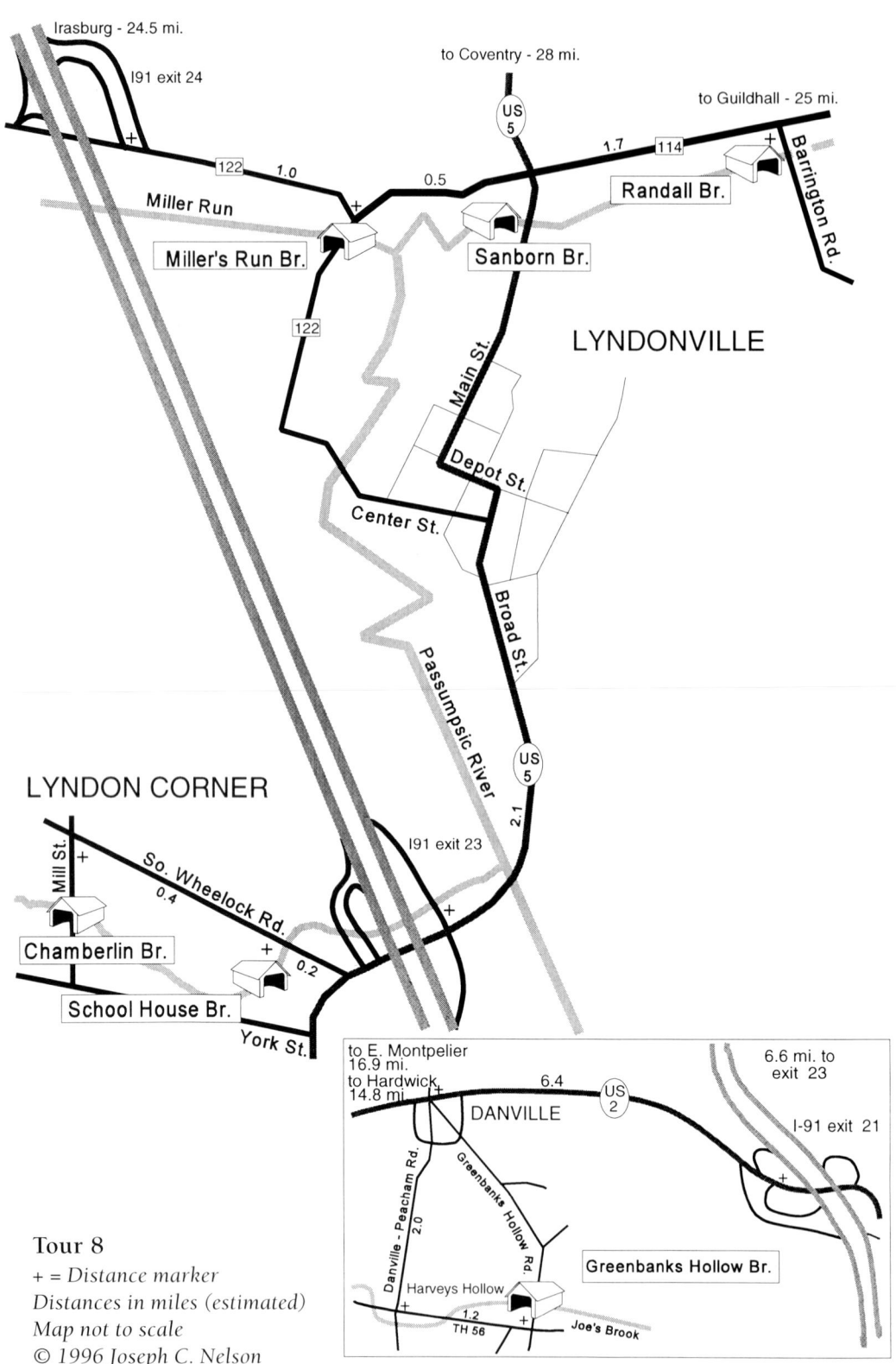

Tour 8
+ = Distance marker
Distances in miles (estimated)
Map not to scale
© 1996 Joseph C. Nelson

The LYNDON BRIDGES

Tour 8

Four of the five bridges in Lyndon have an appearance unique to this region—the "Lyndon look," perhaps. The bridges—Chamberlin, Miller's Run, Randall, and Sanborn—have wide flaring roofs with generous overhangs at the eave and gable ends. The overhangs alone protect the trusses from the weather. The siding, only a few feet high, exposes the interiors to light and air. The visual effect is that of a floating roof. A fifth span, the School House Bridge, contrasts with the other four. It is one of only two bridges in Vermont with the truss completely enclosed, not only on the bridge exterior, but on the roadway side as well.

Except for the Sanborn Bridge, the Lyndon bridges use the queenpost truss. The Sanborn Bridge is one of only three Paddleford truss bridges surviving in the state.

One other bridge is so obviously related to the three sideless queenpost Lyndon bridges that it must be included here: Greenbanks Hollow Bridge in the neighboring town of Danville. These four bridges are Vermont's last examples of open-sided queenpost construction.

Lyndon

Rhode Island Doctor Jonathan Arnold and his associates were granted the town lands by the General Assembly of Vermont. Named for Arnold's eldest son, Josias Lyndon, the town was chartered November 20, 1780.

The Honorable George C. Cahoon described the Lyndon of those years in the *Vermont Historical Gazetteer:* "[It is] interspersed with hills and valleys and carved out by the many tributaries of the Passumpsic[1] forming into one beautiful river, its waters uncommonly cold and pure. Several sites of excellent waterpower for mills and machinery are located [there]."

When Lyndonville was incorporated in 1880, Isaac W. Sanborn, Justice of the Peace, led the organization meeting. Among the officers elected were Isaac Sanborn, town clerk and treasurer, and J. C. Jones, trustee and water commissioner. These gentlemen played important roles in the histories of the town's wooden bridges.

School House Bridge

With interior and exterior painted white, the neat little School House Bridge stands in a carefully maintained green. Built in 1879, the queenpost trusses are completely sheathed with planking, the structure additionally protected by overhanging eaves and the gable-end extensions above the portals. The portals are triple-arched and supported on enclosed corbels finished with pilasters. A pedestrian walkway is tucked under an eave overhang, with the walkway separated from the bridge deck by the truss. Originally, there was a walkway on each side to accommodate the comings and goings of the children from the nearby school, but there is no record of when or why one walkway was removed.

Construction began on the Lyndon Academy and Graded School in 1871. Two new roads provided access to it, and one of them crossed the South Branch to South Wheelock Road. In the October 4 issue, the editor of the *Vermont Union* panned a bridge completed there in 1872: "The most extravagant piece of work we have seen in a long time . . . if it cost anything, the town will be cheated . . . ."

The unpopular bridge didn't last very long. On December 12, 1879, the *Vermont Union* reported: "The new bridge on School Street is completed and is a job well done. John Clement laid the abutments which is a guarantee that the work will stand. J. C. Jones drew the plan of the woodwork and Lee Goodell framed it and super-

> **School House Bridge**
>
> TOWN: Lyndon Corner
> DATE: 1879
> TRUSS LENGTH: 41'8"
> BUILDER: J. C. Jones, Lee Goodell
> TRUSS TYPE: Queenpost

[1] Passumpsic - Algonquian for "Clear waters."

The School House Bridge was the second bridge constructed to serve the Lyndon Academy and Graded School. Curiously, the trusses are each planked over on both sides. Perhaps the authorities had this done to keep school children from climbing up to the roof.

vised the building . . ." (In the July 31, 1896, *Vermont Union*, the editor noted that John Clement had laid the foundations of thirty Lyndon bridges.)

In 1931 John B. Chase wrote in the *Vermont Union Journal*: "The bridge had its hardest test in the 1927 flood when it was tipped up till the chances looked about 100 to 1 that it was going out, but it withstood the onslaught of the rushing water and debris piled against it. As the water went down, the bridge gradually settled back into place, little if any damaged."

In 1971 construction of Interstate Highway 91 required relocation of the road served by the bridge. The owners of a local business donated an acre of land. Refurbished and surrounded by a small park created with local donations, the bridge was placed on the National Register of Historic Places.

The forty-two-foot bridge crosses the South Wheelock Branch of the Passumpsic River off South Wheelock Road. Take Route 5 west from I-91 exit 23 and proceed for 0.2 miles, then turn onto South Wheelock Road. Continue west on South Wheelock Road for 0.4 miles to find Mill Street and the Chamberlin Bridge.

Chamberlin Bridge

The Chamberlin Bridge serves York Street and Mill Street west of Route 5. The most westerly of Lyndon's string of five bridges, the bridge has also been known as the Chamberlin Mill Bridge, the Sawmill Bridge, and the Whitcomb Bridge.

The sixty-six-foot structure spans the North Branch of the Passumpsic River. The original unmortared stone construction is still visible in the north abutment. The south abutment was faced with concrete, probably in the 1960s. The bridge is used daily by light traffic.

The construction features open sides, with the queenpost truss protected by the wide eaves of an overhanging metal roof. The extended gable-ends and the truss are painted white. The gable sheathing, cut high over the roadway, dips to the corbel ends and up again to the ends of the eaves, all in straight lines.

It is not known when the present bridge was built, but it is known that it is not the first on this site. A 1795 map shows a bridge crossing the stream here—it and the Chamberlin Bridge's other predecessors were probably open bridges. The *Vermont Union* newspaper reported in

The Chamberlin Bridge, once an open span, was covered in 1881. Standing high over the South Wheelock Branch of the Passumpsic River, it was one of the few spans not carried away by the great flood of 1927. The white-gabled bridge still serves light traffic.

Chamberlin Bridge

TOWN: Lyndon Corner
DATE: Unknown
TRUSS LENGTH: 65'11"
BUILDER: W. W. Heath (superintendent)
TRUSS TYPE: Queenpost

August 1881: "The Chamberlin Bridge at the west of this village [Lyndon Corner] is having a new abutment and is to be built over into a covered bridge." Selectman W. W. Heath superintended the work.

Ephraim Chamberlin built a gristmill here before 1817 and later added a sawmill. Anson Miller of Dummerston built a wagon and sleigh works at the north end of the bridge in 1818. It was here that the famous Lyndon Cutters were built. These horse-drawn buggies weighed a mere 175 to 275 pounds compared to the usual four hundred pounds or more. In 1840 Myron Chamberlin, Ephraim's son, built a new gristmill. Sharing the water power with the Chamberlins and the Millers in this busy place were a carding mill, a fulling mill, a bark mill, a saddler's shop, and an oil mill.

Through the years the mills changed hands several times, but carried the Chamberlin name until 1905. Harold Whitcomb bought the works at that time and converted them from water power to electricity. When Whitcomb sold out in 1937, the mills and the bridge reverted to the Chamberlin name again.

The great flood of 1927 struck during Whitcomb's time. Luther Johnson commented in his *Vermont in Flood Time*: "At Lyndon village all the bridges along Little York Street, except the one at Whitcomb's Mill, were carried away." The mill the bridge was named for burned in the 1950s. The stone foundations can still be seen.

To view the three bridges to the north, return to Route 5. There, turn left and go north about two miles to the Sanborn Bridge.

Sanborn Bridge

The handsome Sanborn Bridge is typical of the Lyndon bridges in the construction of the roof—there is a generous eave overhang and gable-ends extended on corbels. The gable sheathing is painted red under a sharply pitched roof. The portals are triple-arched and trimmed with white molding, and there is a pedestrian walkway on one side of the main deck.

Sanborn Bridge

TOWN: Lyndon
DATE: 1869[1]
TRUSS LENGTH: 120'
BUILDER: Unknown
TRUSS TYPE: Paddleford

---

[1] While the sign at the gable says 1867, N.G. Templeton dates the Sanborn Bridge 1869 in his pamphlet *Vermont's Covered Bridges*. Milton Graton dated it 1873. Mr. Graton claimed the ability to date a bridge by the characteristics of the timber used in the construction.

The Sanborn Bridge was moved from where it once crossed the Passumpsic River to Benjamin Sanborn's farm. The Sanborn is the last of the many Paddleford truss bridges that once crossed the stretch of river between Lyndon and St. Johnsbury. The others were destroyed in the 1927 flood.

A road and a bridge were constructed on Benjamin Sanborn's meadow in 1858 to connect Lyndon Center with farms on the west side of the Passumpsic. What type of bridge was built there and who the builders were is unknown. The width of the river required a bridge-span of 120 feet, a distance beyond the capabilities of the queenpost truss commonly used in the area, but within the range of Paddleford's truss. Ithiel Town's plank-lattice truss would have qualified, but for some reason the Town truss is not found in this part of Vermont. If there were an interim bridge at this location prior to the dates attributed to the Sanborn Bridge, it could have used a Paddleford truss, a popular design in northeast Vermont and neighboring New Hampshire.

The Sanborn now spans the West Branch of the Passumpsic River a few yards west of Route 5 near the junction of Route 114. In 1960 Milton Graton, covered bridge restorer, moved the bridge through the streets of the village to its present location. Owned by the motel beside it and no longer serving as a highway bridge, it has housed a series of businesses.

To view the Miller's Run and the Randall bridges, continue north on Route 5 to the junction of Route 114. There, turn left and drive 0.5 miles to the Miller's Run Bridge. To view the Randall Bridge, turn right and drive 1.7 miles.

Miller's Run Bridge

The newly reconstructed Miller's Run Bridge is clean and spare with no ornaments. It is painted white inside and out, with the low siding exposing the queenpost truss. The bridge was originally built in 1878 and was a no-nonsense Yankee answer to the need to cross a river. Until it was reconstructed in 1995, the little bridge was in constant use.

The queenpost trusses of the original bridge were incorporated into the reconstruction to support only the roof. A modern steel structure under the wooden deck supports the roadway. The reconstructed bridge is true to the Lyndon look, with its open sides and flaring roof. The fifty-four-foot bridge is still single-lane, but it is a bit wider—the old tie beams were replaced with longer timbers of southern yellow pine. The new span features an attached foot-bridge not found on the original. Also new is the green-enameled standing-rib metal roof. The simple triangular gable ends were replaced with a geometry closer to that of the Chamberlin and Randall bridges.

The bridge recently replaced was not the first to serve here. The Board of Selectmen levied a special tax in 1800 to build a span near "the mount of Miller's Run." What probably was an open bridge was repaired in 1816 and rebuilt in 1841. The first reference to a covered bridge at this site appeared in the *Vermont Union* in August 1878: "The Selectmen have completed a new covered bridge over Miller's Run. It is a new style bridge for these parts and is said to be the best in town." E. H. Stone drew the plans and supervised construction.

The new bridge over Miller's Run continues to serve travelers on Route 122. It is located in the northwest cor-

---

Miller's Run Bridge

TOWN: Lyndon Center
DATE: 1995
TRUSS LENGTH: 53'8"
BUILDER: E. H. Stone (superintendent)
TRUSS TYPE: Plank-lattice

Lauded as a "new style" bridge in 1878, the Miller's Run span was updated in 1995. The original trusses decorate a new bridge that features a wooden deck on steel beams and a pedestrian walkway.

ner of Lyndonville Village, 0.5 miles west of the junction of Routes 5 and 114, and one mile west of I-91 exit 24.

Randall Bridge

The old Randall Bridge is retired. It stands next to Barrington Bridge Road just off Route 114 north of Lyndonville, about 1.7 miles from the junction of Routes 5 and 122. Spanning the East Branch of the Passumpsic River, it serves a footpath and snowmobile trail. In 1965 a new concrete and steel bridge was built here. The wooden bridge was left standing as an example of the craft of the covered bridge builders of a century or more ago.

Like the Miller's Run Bridge and the Chamberlin Bridge, the Randall is constructed with open sides, the queenpost truss protected by the wide eaves of the metal

Randall Bridge

TOWN: Lyndonville
DATE: 1865
TRUSS LENGTH: 67'2"
BUILDER: Unknown
TRUSS TYPE: Queenpost

roof. The extended gable-ends are stained brown. The gable sheathing, like that of the Chamberlin, is cut high over the roadway, dips to the corbels, and extends up again to the ends of the eaves. The interior and truss are unpainted.

Danville

Chartered in 1786, Danville was named for the French Admiral Le Duc D'Anville. During their struggle to form a state in the New Hampshire Grants, the efforts of Ethan Allen and his associates were encouraged by the French Consul at Boston. In appreciation, they named several townships in honor of distinguished Frenchmen.

The selection of D'Anville seems strange, however. A look back in history reveals that Admiral D'Anville led a French mission to attack Boston in retaliation for the 1745 taking of Louisburg on Cape Breton Island by a force of New Englanders. The Admiral's assault was nipped in the bud only by a North Atlantic storm that drowned him and destroyed most of his fleet.

Like all of the towns chartered and settled in those

The Randall Bridge stands preserved as an example of the bridge construction unique to the Lyndon area, with its open sides, wide eaves, and extended gables. Today, the retired bridge is a good place from which to wet a fishing worm.

early years, the economy was based on water power and agriculture. Population centers began where industry could grow, which in Danville was along Joe's Brook. Today, there is very little evidence of those beginnings—except for the covered bridge and the remains of the old mill works, the villages are gone.

Greenbanks Hollow Bridge

Ghostly in its coat of white paint, the fifty-foot span of Greenbanks Hollow Bridge crosses Joe's Brook just upstream of a broken dam. Stone foundations stand dark in the brush along the banks of the stream. A bronze plaque mounted on a stone by the bridge portal reads: "Historic Site Greenbanks Hollow Covered Bridge 1886 Danville."

A bridge built here in the early 1800s burned in 1885 and was rebuilt in 1886. H. W. Congdon in *The Covered Bridge* believes the original bridge was built without a roof.

In 1970 the bridge roadway was reinforced with steel beams installed on the deck and tie-bolted to the bottom chords below. The structure is also supported with piers.

Greenbanks Hollow Bridge crosses Joe's Brook just upstream of a ruined dam. Ghostlike in low light, this white-painted span sparkles in the sun.

> ### Joe's Pond
>
> Joe's Brook flows from Joe's Pond. Joe's Pond was named for Captain Joe, a Nova Scotia Indian—the nearby Molly's Pond was named for his wife. Captain Joe used to hunt and fish around the pond, and at one time he had a camp there. He was a big help to the settlers with their relations with the local Indian tribes.
>
> During the Revolutionary War, Captain Joe fought against the English, hating them for having dispersed his people in Nova Scotia. He and Molly had the distinction of having dined at the table of General George Washington. After the war they settled in Danville on a pension granted by the State of Vermont.

> ### Greenbanks Hollow Bridge
>
> TOWN: Danville
> DATE: 1886
> TRUSS LENGTH: 74'5"
> BUILDER: Unknown
> TRUSS TYPE: Queenpost

The Vermont Agency of Transportation has recommended that the town consider two options: close or relocate the bridge to a preservation site and build a new structure or reconstruct the bridge for moderate traffic. The Miller's Run Bridge in Lyndon offers an example of such reconstruction.

The bridge is best reached by driving south from Route 2 through Danville on the Danville-Peacham road for two miles. Turn right at Harvey's Hollow onto town highway 56 and drive east one mile to the bridge. Once you leave Harvey's Hollow, the roads are unpaved, and travel during mud season can be chancy.

To return to Route 2, you have the option of crossing the bridge and driving north, back to Danville.

Sources

Congdon, Herbert Wheaton. *The Covered Bridge.* Middlebury, VT: Vermont Books, 1973.

Fisher, Harriet. *A Walk Around Lyndon.* Lyndon, VT: The Lyndon Bicentennial Booklet Committee, 1978.

Graton, Milton S. *The Last of the Covered Bridge Builders.* Plymouth, NH: Clifford-Nicol, Inc., 1978

Hemenway, Abby Maria. *The Vermont Historical Gazetteer, Vol. IV.* Burlington, VT: A. M. Hemenway,, 1877.

Johnson, Luther B. *Vermont in Flood Time: November 1927.* Randolph, VT: Roy L. Johnson Co., 1928

Shores, Venila Lovina. *Lyndon, Gem In The Green.* Lyndonville, VT: Town of Lyndon, 1986.

Vermont Agency of Transportation Covered Bridge Study. Prepared for the State of Vermont Agency of Transportation by McFarland-Johnson, Inc., Binghamton, New York, 1995.

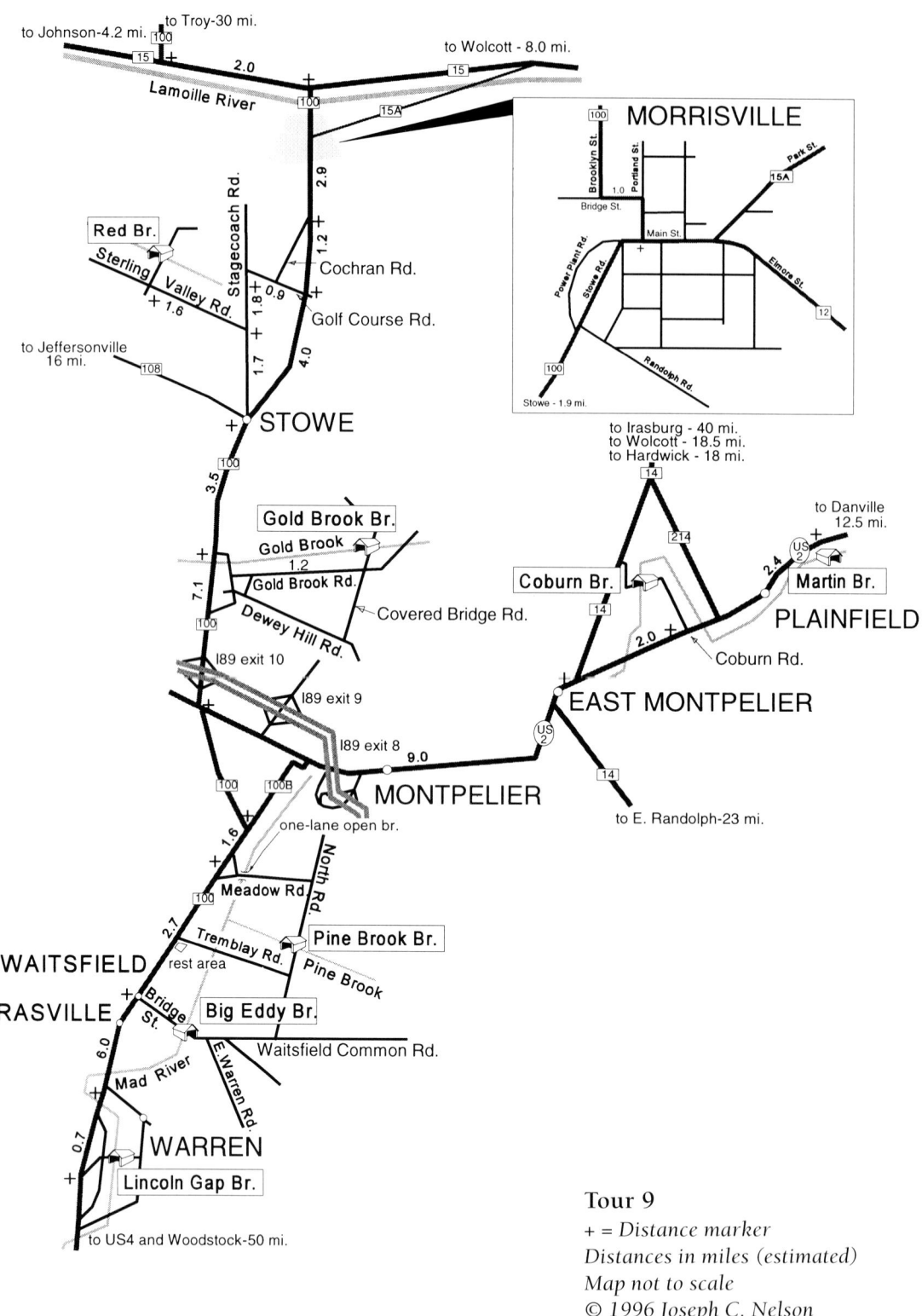

Tour 9
+ = Distance marker
Distances in miles (estimated)
Map not to scale
© 1996 Joseph C. Nelson

ROUTE 100 in CENTRAL VERMONT

Tour 9

There are seven covered bridges of five construction types to be found along the fifty miles of Vermont Route 100 and U.S. Route 2 connecting Marshfield, Morristown, Stowe, and Warren. The scenic setting varies from the mountainside villages of Stowe to the farmland of the Winooski and Mad River valleys.

In the text, the bridge descriptions are laid out from Route 15 and Morristown, south through Stowe to U.S. 2 and Interstate 89. The viewer is guided east on U.S. 2 toward Marshfield, passing the golden statehouse dome in Montpelier along the way. After visiting the bridges in East Montpelier and Marshfield, the bridge seeker returns to Route 100 and heads south to Waitsfield and Warren.

Morrisville/Morristown

Travelers from Route 15 must pass through the streets of Morrisville, either on Route 100 or on Route 15A. The streets are marked with state route signs, so it should be difficult to get lost. Once through Morrisville, follow Route 100 south.

To find the Red Bridge in Morristown, continue south on Route 100 for 1.9 miles and turn right onto Cochran Road. When Cochran Road ends on Golf Course Road, turn right and proceed to Stagecoach Road. There, turn left and go 1.8 miles to Sterling Valley Road. Turn right onto Sterling Valley Road and drive 1.6 miles to the bridge.

Red Bridge

There is a little barn-red covered bridge crossing Sterling Brook in Morristown. It serves Cole Hill Road above a scenic narrow gorge with walls of exposed bedrock. Like most of Vermont's old bridges, it has been known by many names. Residents once referred to it as the Chaffee Bridge for the family who lived near. Others remember it as the Sterling Brook Bridge. Now it is usually called the Red Bridge.

The Red Bridge was built in 1896 using a kingpost truss with a superimposed queenpost truss system. Sixty-four feet long from portal to portal, the little bridge appears to be longer because it has ten additional feet of extended gables. Damaged by a storm in the fall of 1897, the strange truss was made stranger still at that time with the addition of a profusion of iron rods. The resulting truss has been defying classification since.

In 1971 the Vermont Department of Highways reconstructed the bridge with an independent reinforced concrete roadway supported by two steel beams. The original stone abutments were replaced with cast concrete. Though the old truss now supports only itself, the struc-

Red Bridge

TOWN: Morristown
DATE: 1896
TRUSS LENGTH: 64'2"
BUILDER: Unknown
TRUSS TYPE: Unique

Morristown's Red Bridge crosses a scenic gorge carved into the bedrock by Sterling Brook. The span is best known for its unique truss invented by an unknown builder.

ture remains a prime example of nineteenth century bridge tinkering.

Return to Stagecoach Road, turn right, and drive 1.7 miles to Route 100. Continue south 3.5 miles through the Village of Stowe to Gold Brook Road. Follow Gold Brook Road, bearing left at the first intersection. Continue to Covered Bridge Road and the Gold Brook Bridge.

STOWE

Stowe has long been one of Vermont's leading tourist towns. As early as the middle 1800s, it was common for strangers with fine apparel, fine horses, and fine carriages to stay for three or four months of the year. The "summer people" gave the town the reputation of being the Saratoga of Vermont.

Stowe has three villages. Center Village is located at the geographical middle of the town, where all roads meet. A half mile to the south lies Lower Village, once known as Mill Village. This is where the trades, such as the mills and blacksmith shops, were located. The third village is Moscow, two miles south of Center Village. Moscow was known to have one of the best sawmills in the state, as well as a large door, sash, and blind factory.

Gold Brook Bridge

This dark little bridge crosses Gold Brook at a busy joining of well-kept country roads. The impression of darkness comes from the walnut-stained gable-ends, but there is also a dark side to the bridge's history. For this is Emily's bridge, haunted by a locally famous ghost.

Many stories exist about why poor Emily does not rest in peace. Perhaps the most popular legend has it that in the middle 1800s Emily, a farmer's daughter, was deserted by her lover. Despairing and, sadly, in a family way, she hanged herself in the bridge. Her ghost is sometimes seen wandering through the bridge on moonlit midsummer nights, waiting for her man.

The first span to cross Gold Brook at this site was constructed in 1803. John W. Smith designed and built the current Gold Brook Bridge in 1844 using the Howe truss. Finding a Howe truss in a forty-eight-foot highway bridge in the middle of Vermont is something of a surprise—

Gold Brook Bridge

TOWN: Stowe
DATE: 1844
TRUSS LENGTH: 48'5"
BUILDER: John W. Smith
TRUSS TYPE: Howe

The builder of the Gold Brook Bridge was very much at the leading edge of technology when he used the newly patented Howe truss. Howe's design became the standard for railroad bridges until it was superseded by steel trusses.

the design was meant for the much more stringent loading requirements of a railroad span. The truss has historical significance for its use of iron rods and angle blocks. The only other surviving examples of the Howe truss in Vermont are the Connecticut River bridges at Lunenburg and Lemington and the Rutland Railroad Bridge at Shoreham.

Accounts of the Gold Brook Bridge recall that the original timber beams in the deck system were replaced with railroad rails to support the timber roadway. Today the rails are no longer in evidence.

The Gold Brook Bridge is in very good condition and is likely to remain so. In 1969 the town made a resolution for perpetual care. The declaration is displayed on a bronze plaque placed in a grassy area near the east portal.

To continue the tour to the Coburn Bridge, return to Route 100 and turn left to continue south eighteen miles to U.S. Route 2 in Waterbury. Take Route 2 east nine miles, through Duxbury and Montpelier, to East Montpelier and the junction with Route 14. Continue two miles on Route 2 to Coburn Road. Turn left onto Coburn Road

> **Gold in the Brook**
>
> Gold Brook does indeed have some gold in it. In May 1857 Captain A. H. Slayton, who had experience in the California gold fields, discovered some particles of gold on the banks of the brook on the farm owned by Nathaniel Russel. Captain Slayton purchased the farm and hired three or four diggers for several days. The effort did not pay off in the end, but Slayton was able to make a watch chain worth about one hundred 1857 dollars with the gold that was found.

and drive 0.5 miles to the bridge. Coburn Road is narrow and winding with many blind curves, so use care while driving on it.

East Montpelier

The land in what is now East Montpelier was settled in 1788, but it was originally part of the Town of Montpelier. East Montpelier was organized in January 1849, having been split off from Montpelier by the Vermont General Assembly in 1848.

Coburn Bridge

> **Coburn Bridge**
>
> Town: East Montpelier
> Date: 1851
> Truss length: 50'
> Builder: Larned Coburn
> Truss type: Queenpost

The Coburn Bridge is not just a covered bridge, it is a fine representative of the bridge-builder's art. The eaves are wide, shedding rain and snow far away from the structure. The gable ends are deep for the same purpose. The sheathing on the sides is three-quarter height, letting in light and air. These are the features that craftsmen employed to ensure a long life for their bridges. But the pilasters and trim at the portals serve no practical purpose—these finishing touches are marks of the builder's pride.

Built in 1851 by Larned Coburn, the fifty-foot structure spans the Winooski River perched high on its abutments, out of reach of the last freshet and hopefully out of reach of the next. The Coburns, as long-time residents, were familiar with high water in this place. The patriarch, Larned Coburn, settled in East Montpelier in 1830. The Coburn homestead, the land adjacent to the bridge, is celebrated as having remained in the same family for more than one hundred years. Luther Johnson, in his

Vermont in Flood Time: November 1927, tells a story of Larned's descendants during the great flood:

"The eight cows of F. W. Coburn were drowned when the Winooski surrounded his farm buildings at the Plainsfield, East Montpelier line. A rescue party which included Mr. Coburn's son Harry L. Coburn, tried to reach the dwelling on a raft but it capsized. The party managed to gain a tree from which Mr. Coburn drew them to the house by means of a rope, rescuing the rescuers."

The bridge was reconstructed in 1972 by the Agency of Transportation. A concrete deck on steel beams now supports traffic, and the old truss supports only itself. Despite the new work done, the basic technology of the original work is still evident. The upper chords were fitted into place mostly "in the round," and the diagonals and cross-beams clearly show the marks of the adz and broad axe. The truss itself was renovated in the winter of 1996–97, so new carpentry is also in evidence.

Return to Route 2 and turn left. Martin Bridge stands in sight of Route 2, three miles past Coburn Road. Watch to the south as you approach a barn and a row of small houses on the left side of the highway.

Located near a gravel quarry, the Coburn Bridge's queenpost truss was tested by heavy loads for many years. The wooden deck was replaced in 1972 with a self-supporting concrete slab. Unique for its hand-worked timbers, the old truss was renovated in the winter of 1996–97.

Marshfield

The Town of Marshfield was granted to the Stockbridge Indian Tribe by the General Assembly of Vermont in 1790. The Indians intended to settle here, but after white settlements were founded around their town, they sold it to Captain Isaac Marsh of Stockbridge, Massachusetts. The Indians moved on to the then unsettled forests of New York.

William Martin was an early settler of Marshfield. He bought a farm about a mile north of Plainfield Village and resided there until 1840. His farm was reputed to be one of the finest on the headwaters of the Winooski River. The Ortons bought the old Martin place and gave it their name for a time.

Martin Bridge

The Martin Bridge, or Orton Farm Bridge, crosses the Winooski River in a pasture south of Route 2, three miles east of the Coburn Road intersection. Built in 1890, it is believed by some to be the last surviving example of the work of Herman F. Townsend.

The forty-five-foot queenpost truss structure is privately owned. It stands high on abutments of cut granite and rubble stone laid dry. There is no road leading to the bridge and no ramps to permit a vehicle to enter. The floor planking is gapped and unsound. A cattle gate is hinged at one of the queenposts.

To view the bridges in Waitsfield and Warren, follow Route 2 back to Duxbury. There, take Route 100B south to the junction of Route 100. Continue south 1.6 miles on Route 100. Turn left on Meadow Road and cross an open one-lane bridge. Continue on Meadow Road to North Road. Turn right and drive to the Pine Brook Bridge.

Martin Bridge

TOWN: Marshfield
DATE: 1890
TRUSS LENGTH: 44'9"
BUILDER: Herman F. Townsend
TRUSS TYPE: Queenpost

Waitsfield

The Waitsfield Town charter was signed in 1782 by Governor Thomas B. Chittenden. Waitsfield was named for its first settler, General Benjamin Wait, who was one of the New World's original warriors. Wait fought the French in the early days of the colonies, then against the British

The Martin Bridge is named for a pioneering Marshfield settler who farmed this land. It stands in alluvial fields watered by the Winooski River.

in the Revolutionary War. He also worked against New York interests in the formative years of the New Hampshire Grants, which were to become Vermont.

Construction of roads and bridges began soon after the town was chartered. In 1806, however, all of the bridges on the Mad River were damaged by a flood. Floods destroyed all of the bridges again in 1824, then one more time in 1830. After that, the voters finally realized that they needed sturdier, more permanent structures.

Pine Brook Bridge

The Pine Brook Bridge, or Wilder Bridge, is a fine example of classic kingpost truss construction. It is one of Vermont's four surviving kingpost bridges and one of two that actually features wooden kingposts. The little bridge,

> **Pine Brook Bridge**
>
> TOWN: Waitsfield
> DATE: 1872
> TRUSS LENGTH: 48'7"
> BUILDER: Unknown
> TRUSS TYPE: Kingpost

There were once hundreds of small kingpost spans like the Pine Brook Bridge. In Vermont, all but four have been replaced with culverts and concrete bridges.

built in 1872, was carefully and authentically restored in 1976 by Milton Graton and Sons. The forty-nine-foot span serves North Road just one mile north of Waitsfield Common.

The bridge was in a poor state when Graton inspected it. The chords had rotted, a corner of the bridge had sagged almost a foot, and there were posts standing in the stream supporting the middle of the span.

Graton, in his book *The Last of the Covered Bridge Builders*, relates that the restoration had to conform to the requirements of historic preservation to qualify for federal aid, so setting the bridge on a concrete pad and supporting it with steel beams would not do. He proposed setting the steel beams under the bridge one-half inch short of contacting the chords as a safety device for heavy loads.

The bridge was raised a foot and a half to keep the deck system dry. The unmortared stone abutments were capped, the steel beams were installed, the chords and siding were replaced, and the roof was repaired. The floor was re-decked in 1989.

ROUTE 100 IN CENTRAL VERMONT

Drive through the bridge and continue to Tremblay Road. There, turn right to return to Route 100. Continue south on Route 100 to Waitsfield. Turn left onto Bridge Street to cross the bridge, or, to get the best view, continue a little farther on Route 100 to find the riverside parking lot off to the left.

Big Eddy Bridge

The Big Eddy Bridge was built in 1833 to replace a span lost in the freshet of 1830. It is thought to be the second oldest Burr-arch-type structure in Vermont after the Pulp Mill Bridge in Weybridge-Middlebury, which is believed to have been completed in 1820.

Big Eddy stands in the middle of Waitsfield Village and stretches 105 feet over the Mad River, serving Bridge Street and East Warren Road. After 140 years of service, the old bridge was a wreck, wounded by years of pounding by the elements and heavy use, so, in 1973, Milton Graton and Sons of New Hampshire contracted to return the bridge to good health. It stands today ready for another 140 years.

Graton replaced the deck structure, flooring, and the upper lateral bracing. He also installed new knee braces, hewn from the branches of trees, and re-sheathed the roof with standing seam metal sheeting. The foot bridge, added to the side of the span in 1940, was restored and reattached.

In 1989 the floor was re-decked and distribution beams were added under the deck. In 1992 Paul Ide and Jan Lewandoski made structural repairs to the truss. All of the work has been done using materials and methods that maintain historic authenticity, leaving the basic original structure intact.

It is interesting to cross the bridge on the walkway and look down at the Mad River. The bedrock outcrop supporting the north abutment juts into the current, creating the swirl that gives the bridge its name. The bridge's surroundings are very picturesque and capture the essence of an old-fashioned New England village.

To complete the tour, continue south six miles on Route 100 to Warren Village.

Big Eddy Bridge

Town: Waitsfield
Date: 1833
Truss length: 105'1"
Builder: Unknown
Truss type: Burr arch

Built primarily to withstand the Mad River at its maddest, the Big Eddy Bridge was built without frills soon after the freshet of 1830 cleared the valley of bridges. A pedestrian walkway was added later.

Warren

The Town of Warren was chartered to John Throop and associates on November 9, 1780. With the Mad River running through the center of the town, Warren developed as a mill town. The town hosted a hoe handle factory, a clothes pin factory, a dowel, chair stock, and rolling pin mill, and the Warren Wooden Bowl factory. Walter Bagley, the builder of the Lincoln Gap Bridge, manufactured clapboards in South Hollow for several years.

Lincoln Gap Bridge

The Lincoln Gap Bridge crosses the Mad River just east of Route 100 at Warren Village, about one mile north of west-bound Lincoln Gap Road. The sixty-five-foot structure, ten feet of it gable-end overhang, is one of two queenpost bridges in Vermont in which the truss inte-

The Lincoln Gap Bridge stands in the Historic Residential District in Warren Village, surrounded by signs of nineteenth century industry.

rior and exterior are completely enclosed—the other is the School House Bridge in Lyndon.

Walter Bagley built the Lincoln Gap bridge in 1880. Bagley's claim to fame may be the peculiar construction of the bridge he left us. The roadway was strengthened with timber cross-braces and stringers and the roof was re-shingled in the fall of 1995 by Paul Ide and Jan Lewandoski.

The bridge is located within the village Historic Residential District. The village stands in the folds of a jumble of short, steep hillocks, displaying a mix of old New England and Victorian architecture. The old mill dam stands as a ruin downstream from the bridge.

Lincoln Gap Bridge

TOWN: Warren
DATE: 1880
TRUSS LENGTH: 54'11"
BUILDER: Walter Bagley
TRUSS TYPE: Plank-lattice

SOURCES

East Montpelier Bicentennial Souvenir Booklet, 1976.

Graton, Milton S. *The Last of the Covered Bridge Builders.* Plymouth, NH: Clifford-Nicol, Inc., 1978

Hagerman, Robert L. *Covered Bridges of Lamoille County.* Essex Junction, VT: Essex Publishing Co., 1972

Hemenway, Abby Maria. *The Vermont Historical Gazetteer, Vol. II.* Burlington, VT: A. M. Hemenway, 1871.

Hemenway, Abby Maria. *The Vermont Historical Gazetteer, Vol. IV.* Burlington, VT: A. M. Hemenway, 1877.

Hemenway, Abby Maria. *The Vermont Historical Gazetteer: A History of Washington County.* 1882.

Johnson, Luther B. *Vermont in Flood Time: November 1927.* Randoph, VT: Roy L. Johnson Co., 1928

Lewandoski, Jan. *Wood Truss Highway Bridges in North America: Repair and Strengthening.* Proceedings of the Fifth International Conference on Structural Faults and Repair, pp. 217-223, Edinburgh, Scotland, 1993.

Vermont Agency of Transportation Covered Bridge Study. Prepared for the State of Vermont Agency of Transportation by McFarland-Johnson, Inc., Binghamton, New York, 1995.

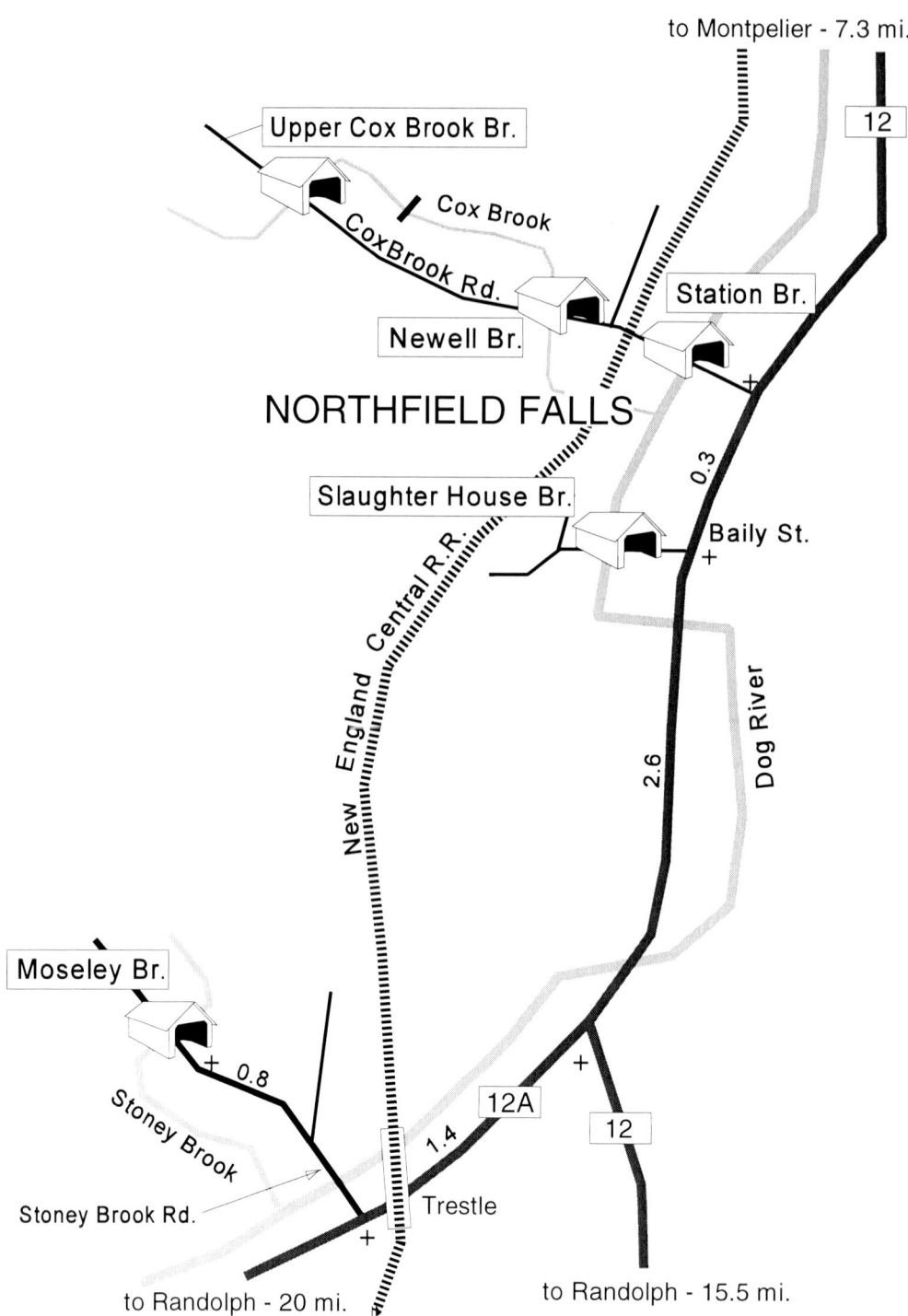

Tour 10
+ = *Distance marker*
Distances in miles (estimated)
Map not to scale
© *1996 Joseph C. Nelson*

NORTHFIELD

Tour 10

Northfield was chartered by the state in 1781, and the first settlers arrived in 1785. According to the Northfield Bicentennial Committee in the pamphlet *Northfield in the Bicentennial Year 1976*, the town was often referred to as "Northfield-on-the-Dog."

The "Dog" is the Dog River. It and its tributaries were the source of the wealth of waterpower important to the early growth of the community. According to legend, the river got its name when a hunter's dog, pursuing a moose, fell through thin ice on the river and drowned.

Northfield began as a small industrial center in the days of waterpower. Elijah Paine founded the first millworks on Robinson Brook at Mill Hill near the village of South Northfield. Other enterprises established themselves along the tributaries of the Dog River, prospered for a time, and then petered out.

On Sunny Brook there was a shingle mill, a carriage shop, a blacksmith shop, a knife factory, a manufacturer of doors and sashes, a chair factory, a gristmill, a wool carding mill, and a manufacturer of coffins and caskets. Mills also operated on Cox Brook, Union Brook, and Stoney Brook. The East Branch of the Dog River, now Bull Run, supported a gristmill, a carriage shop, a machine shop, and a manufacturer of pumps, chairs, and handles for hay forks and brooms. Today, Norwich University is the primary employer. The mills are gone, and the town is primarily a forest area with some crop and pasture land. Five covered bridges tie the area together, as they did in the mill town years.

Northfield Falls, called Gouldsville in the mill days, is located where Cox Brook joins the Dog River. The Station Bridge is the centerpiece of the village, easily seen to the west from Route 12. The portal beyond the Dog River is only a few feet from the New England Central Railroad right-of-way. The Newell Bridge can be seen through the Station Bridge on Cox Brook Road, directly across the railroad tracks. Northfield Falls is the last place in Vermont where you can still view one covered bridge from another. The Upper Bridge is a tenth of a mile further up Cox Brook Road, beyond the Newell, past a dam and on the summit of a low hill.

Four ports on each side of the Station Bridge give travelers the opportunity to watch for oncoming trains. A fine view of the bridge can be had from the rocks upstream.

Slaughter House Bridge serves an unpaved dead-end road about one-quarter mile south of the Station Bridge, east off Route 12. The Moseley Bridge is 4.8 miles south on Stony Brook Road off Route 12A.

The busiest of the village's covered bridges were reconstructed in the 1960s to handle modern traffic. The original deck systems were replaced with independent timber roadways on steel beams. Except for the Slaughter House Bridge, which has not been altered, the old trusses support only themselves and snow load. All of Northfield's covered bridges were painted red in 1978, when the town spruced them up by replacing and repairing the siding.

Station Bridge

The Station Bridge, also known as the Northfield Falls Bridge, is a reminder of less hurried days and serves Cox Brook Road and Northfield Falls with a single lane. A sign over the portal proclaims "Town Lattice 1872." The fluorescent lamps lighting the interior are a concession to modern times. Four ports in each side expose the truss and afford the traveler an opportunity to watch for trains while crossing the bridge to the tracks. New England Central trains pass here regularly.

When the 137-foot plank-lattice structure was strengthened and the original deck system replaced in 1963, a concrete pier was constructed under the center of the span. The original un-mortared stone abutments remain, but they have cast concrete caps. Tie rods extend between the top chords for lateral bracing. Steel cables give additional reinforcement.

Station Bridge

TOWN: Northfield
DATE: 1872
TRUSS LENGTH: 136'10"
BUILDER: Unknown
TRUSS TYPE: Plank-lattice

Newell Bridge

The Newell Bridge, also known as the Lower Cox Brook Bridge, crosses Cox Brook a short distance from where the stream passes under the railway to empty into the Dog River. The original stone abutments supporting the fifty-six-foot span were faced with concrete when the roadway was strengthened in the 1960s.

A sign over the board and batten portal proclaims "1872 Queen post." A rare cast iron sign spells out "Speed

---

Newell Bridge

TOWN: Northfield
DATE: 1872
TRUSS LENGTH: 56'5"
BUILDER: Unknown
TRUSS TYPE: Queenpost

The Newell Bridge is sometimes called the Middle Bridge because of its position—it is the only covered bridge in the state where one can turn and view another.

limit horses at a walk motor vehicles 10 miles per hour." Another sign forbids loitering. A small port on the south side was created by pushing part of the vertical siding out a few inches.

Upper Cox Brook Bridge

The Upper Cox Brook Bridge stands above a waterfall and an abandoned dam a few hundred yards up Cox Brook Road from the Newell Bridge. The place is seldom lonely in summer—the pool above the falls is a popular swimming hole with lots of places to sit on the bedrock in the shadow of the span.

The Division of Historic Sites, in applying to put the bridge on the National Register of Historic Places, states that the date of build is unknown. The Agency of Transportation, in its bridge inspection report, guesses the date to be about 1872.

The Upper Bridge is vented on both sides for most of its fifty-one-foot length, as the siding is a couple of feet short of reaching the eaves. In addition to the eave vents, there are ports on each side, both created by pushing the

Upper Cox Brook Bridge

TOWN: Northfield
DATE: c. 1872
TRUSS LENGTH: 51'3"
BUILDER: Unknown
TRUSS TYPE: Queenpost

VERMONT'S COVERED BRIDGES

The Upper Cox Brook Bridge keeps its trusses dry with circulating air from eave vents and long side ports. In addition, the siding is lifted like an awning to shed rain away from the trusses, a system unique to Northfield.

siding away from the truss like an awning. The siding below the ports is also pushed away a bit, directing rainwater away from the main stringers.

The upper chord of the queenpost truss is only half the height of the sides. This might ordinarily suggest the span was once open and the was roof added later, except that the queenposts themselves are full height. Because the bridge does not cross the stream squarely, the whole structure is skewed. The roadbed was strengthened in 1966.

Slaughter House Bridge

The Slaughter House Bridge, probably built in the late 1860s or early 1870s, stands near the abandoned site of—you guessed it—an old slaughterhouse. The sixty-foot queenpost structure crosses the Dog River on abutments of unmortared stone blocks and slabs above a mill pond and waterfall. Painted barn-red like the others, it is unique in Northfield in that it alone has rounded portals and no side ports. An adjacent grassy field offers easy access to the riverside.

Slaughter House Bridge

TOWN: Northfield
DATE: c. 1872
TRUSS LENGTH: 59'7"
BUILDER: Unknown
TRUSS TYPE: Queenpost

One of the best examples of masonry laid dry can be found in the abutments supporting the Slaughter House Bridge over the Dog River.

While the town's records do not offer a build date, the Agency of Transportation covered bridge inspection report suggests 1872.

Moseley Bridge

Stoney Brook runs over bare bedrock under the bridge built by John Moseley in 1899. The thirty-seven-foot kingpost truss structure is remarkable in that it actually has a wooden kingpost. Those in most of Vermont's other kingpost bridges have been replaced with, constructed with, or augmented by iron rods. The abutments, once known for their large granite blocks, were faced with concrete in 1990. Five steel beams were installed under the deck in 1971.

Stoney Brook Road leaves Route 12A to the west, 1.4 miles south of the junction of Route 12 in Northfield Center. Drive 0.8 miles to the covered bridge.

Moseley Bridge

TOWN: Northfield
DATE: 1899
TRUSS LENGTH: 36'6"
BUILDER: John Moseley
TRUSS TYPE: Kingpost

John Moseley's bridge stands over Stoney Brook in a pine grove. It is one of the four kingpost bridges left in the state, and one of the two that employ a timber kingpost.

SOURCES

Allen, Richard Sanders. *Covered Bridges of the Northeast.* Brattleboro, VT: Stephen Greene Press, 1957.

Hemenway, Abby Maria. *The Vermont Historical Gazetteer: A History of Washington County,* 1882.

Northfield Bicentennial Committee. *Northfield in the Bicentennial Year 1976.* Northfield, VT. 1976

Vermont Agency of Transportation Covered Bridge Study. Prepared for the State of Vermont Agency of Transportation by McFarland-Johnson, Inc., Binghamton, New York, 1995.

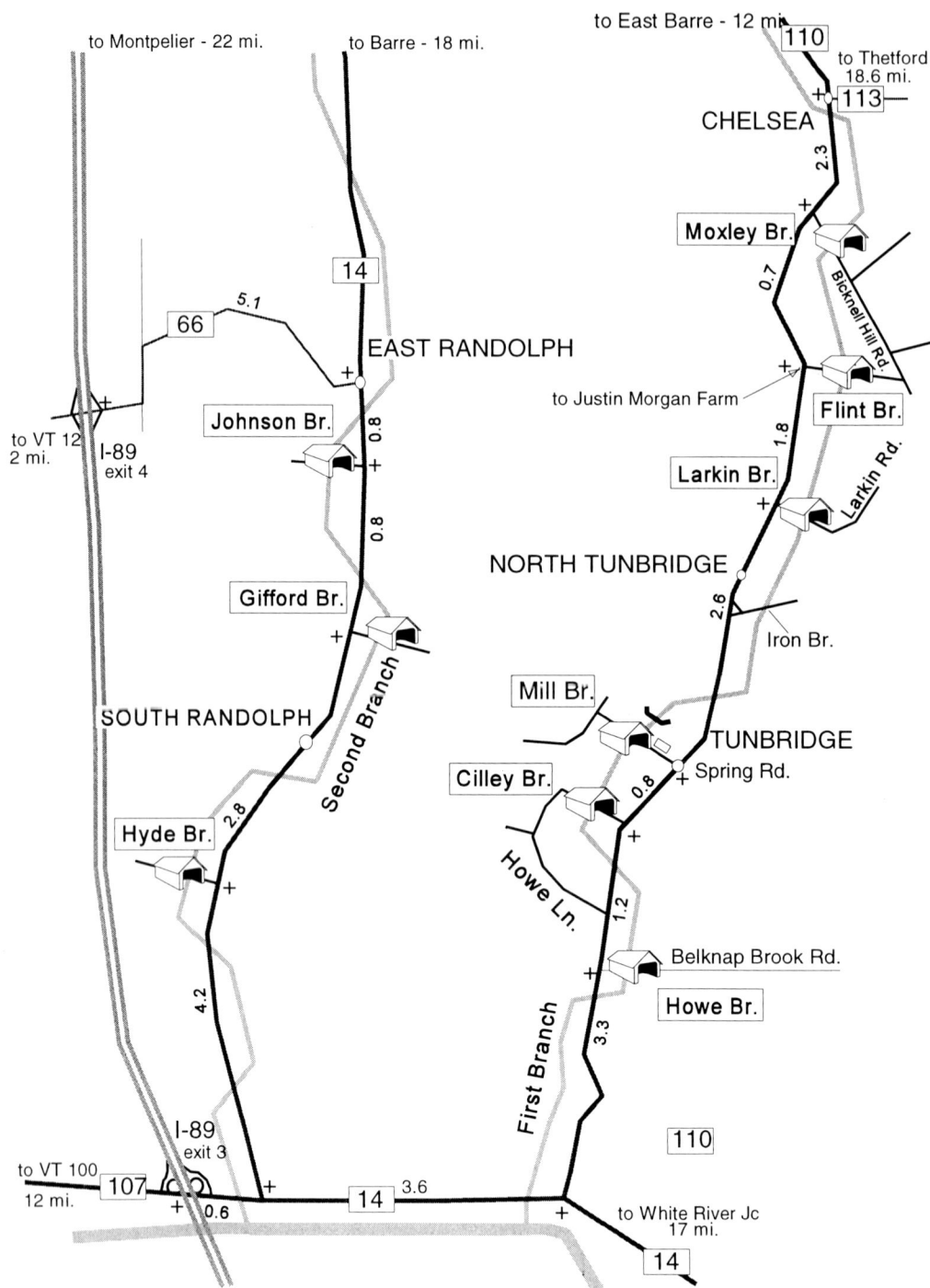

Tour 11
+ = Distance marker
Distances in miles (estimated)
Map not to scale
© 1996 Joseph C. Nelson

The NORTHERN TRIBUTARIES of the WHITE RIVER

Tour 11

The First and Second branches of the White River are easily explored by means of Route 14 and Route 110. Route 14 follows the Second Branch south from East Randolph to South Royalton and the White River. Here, Route 110 leaves Route 14 heading north, tracing the First Branch to Chelsea. The villages scattered along the way represent an older Vermont. The home of the Morgan horse is here, and Tunbridge is host to the famous and frolicsome Tunbridge World's Fair.

Richard Sanders Allen wrote in his *Covered Bridges of the Northeast* that fifteen bridges once crossed the First Branch of the White River, and nine crossed the Second Branch. Of the twenty-four, nine covered bridges survive in the towns of Randolph, Tunbridge, and Chelsea. Two unusual truss types are represented—the Gifford and Johnson bridges in Randolph are the only examples of the multiple-kingpost half-high truss anywhere in the state, and five of Vermont's eight full-height multiple-kingpost bridges are here: Randolph's Hyde Bridge and Tunbridge's Cilley Bridge, Howe Bridge, Larkin Bridge, and Mill Bridge. Tunbridge's Flint Bridge and Chelsea's Moxley Bridge use the queenpost truss.

RANDOLPH AND THE STATE CAPITAL

Randolph was named for Peyton Randolph, president of the Continental Congress of 1774, a leading statesman and friend of George Washington. Since Randolph is located near the geographical center of the state, residents aspired to have the town become the state capital—until 1806, the General Assembly met in various places around the state. Randolph citizens even chose a site for the statehouse and laid out the main street of Randolph Center ten rods wide (165 feet). A mansion was built for the governor.

In 1805 the General Assembly appointed a committee of nine to choose one of three sites: Burlington, Newbury, or Randolph. The group deadlocked, so a compromise had to be found, and they decided to choose a site equidistant from the three towns. This is why the capital city of Vermont, later named Montpelier, was built in the middle of beaver flats next to the Winooski River in Washington County.

Johnson Bridge

The Johnson Bridge, or Braley or Upper Blaisdell Bridge, is forty feet long including the gable-ends that project a protective four feet each. Clean and trim in its red paint, a handmade sign warns "No Trucks or Bus-

---

The Multiple-Kingpost Half-Truss

The first two bridges south of the village of East Randolph on Route 14 are the Johnson Bridge and the Gifford Bridge, both peculiar for their multiple-kingpost half-truss.

The multiple-kingpost half-truss was one of several simple trusses originally used in small open-work bridges. There were once hundreds of open wooden bridges in Vermont, but they are seldom seen today except in private use. All of those serving public roads were ultimately replaced with culverts or concrete and steel, many of them by the bridge replacement program following the flood of 1927.

Some open bridges were simple short-span stringer affairs with no above-deck structure other than railings. Others, longer in span, used kingpost or queenpost trusses. The trusses of the latter were often planked over to preserve them, leaving the deck between them open to the elements. An open deck was not necessarily a bad thing. The decks even in covered bridges had to be replaced regularly due to damage from the dirt and moisture carried in, and by the practice of "snowing the bridge" each winter for the passage of sleighs and sledges. With an open bridge, the town didn't have to hire someone to do the snowing.

Roofs were added to some of them. The Gifford Bridge and the Blaisdell Bridge in Randolph are examples of this. They use a half-high multiple-kingpost truss to which additional structure was added to gain the needed roof-height.

The Johnson Bridge, one of two surviving multiple-kingpost half-high truss spans in Vermont, began life as an open bridge.

Johnson Bridge

TOWN: Randolph
DATE: 1904
TRUSS LENGTH: 37'3"
BUILDER: Unknown
TRUSS TYPE: Multiple kingpost half-high

ses." The portals are well done in a simple and clean style—the gable sheathing is trimmed with a radius that neatly meets the supporting corbels. H. W. Congdon wrote in *The Covered Bridge* that the roof was added in 1909.

Notice that the main stringers have been strengthened with iron rods hung from the added upper chords. Steel cables stabilize the upper works. The roadway was reinforced in 1977 with four steel beams placed under the timber deck.

The Johnson Bridge, out of view from the highway, serves a narrow dead-end dirt road that is easily missed. Look for a sign indicating a road leaving Route 14 to the west about 0.8 miles south of the junction of Route 66 in East Randolph. A barn with a single silo stands across the road from the turnoff. Watch out for the goat!

Gifford Bridge

The Gifford Bridge is outwardly similar to the Johnson Bridge, but it is just a little longer. Also known as the C. K. Smith Bridge, it is sixty feet long, including gable overhangs of five feet each. Years ago the span was called the Blue Bridge, having once been painted that startling color.

Built atop the Gifford's multiple-kingpost half-high truss is another truss resembling a half-high queenpost. It provides the sides with the height needed for a roof and serves to strengthen the span. Iron rods extend between the two top chords for lateral bracing. Here too, as with the Johnson Bridge, the structure is stabilized with steel cables. Two large steel beams laid atop the deck and each side of the roadway support smaller steel beams under the timber floor with tie rods.

Because of suspected use by overweight vehicles and lack of a suitable detour, the Agency of Transportation is recommending that a bypass be built and the bridge closed.

The Gifford, 0.8 miles south of the Johnson Bridge, is easily seen from the highway. A narrow dirt road leaves Route 14 to the east just south of a small concrete bridge by a house with a sign that reads: "The Gifford Homestead Farm."

The Gifford Bridge is constructed with a multiple-kingpost, half-high truss, indicating that it, like the nearby Johnson Bridge, was originally an open span. Notice that the roadway is reinforced with steel beams tie-bolted below the deck.

Gifford Bridge

TOWN: Randolph
DATE: 1904
TRUSS LENGTH: 51'8"
BUILDER: Unknown
TRUSS TYPE: Multiple kingpost half-high

Hyde Bridge

The gable end of the Hyde Bridge, or Kingsbury Bridge, is marked by a large sign that says "So. Randolph, VT." The Hyde alone among the nine area bridges features rounded portals. The span was built in 1904 and restored

> **Hyde Bridge**
>
> TOWN: Randolph
> DATE: 1904
> TRUSS LENGTH: 51'8"
> BUILDER: Unknown
> TRUSS TYPE: Multiple kingpost

in 1980. Since then, it was closed to traffic for a number of years because of ice damage. The damage did not deter a farmer, who used the closed bridge for storing his equipment out of the weather. It was reopened in 1994 after reconstruction. Portions of the top and bottom chords, several vertical truss-members, and some of the bracing was replaced. A bearing block and some bolster beams were renewed.

The fifty-two-foot multiple-kingpost span stands next to Route 14, 2.8 miles south of the Gifford Bridge behind a grassy park with a picnic table. The Hyde serves an unpaved dead-end road, providing access to fields and meadow land.

TUNBRIDGE AND THE WHITE RIVER, FIRST BRANCH

Tunbridge was chartered by Benning Wentworth in February 3, 1761, to Abner Root, Obadiah Noble, and others. The town was named for the Viscount of Tunbridge in England—William Henry Zulestein de Nassau.

Route 110 leaves Route 14 and the White River to head north, following the First Branch upstream to Chelsea,

The Hyde Bridge serves a rural unpaved road next to Route 14. It is now open to traffic after undergoing extensive repairs to fix ice damage.

passing six wooden bridges on the way. Notice that three of the bridges are decorated with flowers, potted and placed in escutcheon-mounted buckets each spring, evidence that Tunbridge cares about its bridges.

Howe Bridge

The Howe Bridge, neatly kept and decorated with flower pots on each side of the portal, serves not only Belknap Brook Road but as an entry to the Howe family farm. The seventy-four-foot multiple-kingpost structure is attributed to Ira Mudget, Edward Wells, and Chauncey Tenny. The deck, floor beams, and bottom chords were replaced in 1994.

In the interest of preservation, the Agency of Transportation recommends that the bridge be closed and bypassed. Agency inspectors cited the presence of the working farm as an indication of the likelihood of use by vehicles exceeding the capacity of the wooden bridge. Alternatively, the old bridge could be rehabilitated to support heavier traffic.

Visitors should be prepared to park at a distance and walk to the bridge as parking in the area is very limited.

Howe Bridge

TOWN: Tunbridge
DATE: 1879
TRUSS LENGTH: 74'4"
BUILDER: Mudget, Wells, Tenny
TRUSS TYPE: Plank-lattice

The Howe Bridge serves as a gateway to the Howe family farm. The portals are decorated with flowers set in antique cattle feeding buckets.

The Morgan Horse

Justin Morgan, his wife, and their three children came to Randolph in 1788. Morgan was a school teacher and singing master, and, for a term, a town clerk. He and his family never owned property, so exactly where the Morgans resided is not known. He is remembered for the horse he brought with him, acquired in the payment of a debt. Then named Figure, the horse became the foundation sire of a breed of horses now known as the Morgan. The Morgan horse is remarkable for its strength, intelligence, and speed.

The Howe Bridge stands at the east median of Route 110, 3.3 miles from the junction with Route 14. A sign at each gable proclaims "Howe Bridge 1879."

Cilley Bridge

The Cilley Bridge, or Lower Bridge, serves Howe Lane just south of Tunbridge—can be viewed from Route 110 to the east across open pasture land. There are signboards on each gable saying "Cilley Bridge - 1883." Its multiple-

The Cilley Bridge crosses the First Branch of the White River at the edge of the river bottom lands. The hillside is covered with brightly flowering ground cover where it rises sharply at the bridge portal.

THE NORTHERN TRIBUTARIES OF THE WHITE RIVER

kingpost truss is sixty feet long, and there is a port on the south side to give users a view of oncoming traffic past a sharp bend in the road. A steep bank at the bridge portal is planted with purple vetch.

Howe Lane leaves Route 110 just 1.2 miles north of the Howe Bridge. Pass two old cemeteries to reach the bridge site, which is located next to a cornfield. Continue through the bridge, bearing left, to come back to Route 110 on Howe Lane. Travel on these unpaved back roads is not advised for the casual driver during winter or mud season.

Cilley Bridge	
Town:	Tunbridge
Date:	1883
Truss length:	59'10"
Builder:	Unknown
Truss type:	Multiple kingpost

Mill Bridge

The Mill Bridge serves a busy road in the middle of Tunbridge, just west of Route 110. Name boards on each gable proclaim "Mill Bridge - 1883," and another sign promises a "One dollar fine for a person to drive a horse or other beast faster than a walk or drive more than one loaded team at the same time on this bridge."

The Mill Bridge stands over the First Branch of the White River in the middle of Tunbridge Village. The dam that ran the mill is a few yards upstream, and the Tunbridge Fairground is close by.

> **Tunbridge World's Fair**
>
> "Looked forward to by most people, dreaded by many," wrote Euclid Farnham in his A *Pictorial History of Tunbridge*. "The fair for one week changes a town of 900 to one of 30,000." The Tunbridge World's Fair was started by the Union Agricultural Society of North Tunbridge in 1867. It was moved to its present site, south of the Mill Bridge, in 1875.

> **Mill Bridge**
>
> TOWN: Tunbridge
> DATE: 1883
> TRUSS LENGTH: 72'
> BUILDER: Unknown
> TRUSS TYPE: Multiple kingpost

The bridge, also known as the Hayward and Noble Bridge, uses the multiple-kingpost truss. The dam belonging to the mill that named the seventy-two-foot structure still stands upstream. The waters of the First Branch spill over the dam, flow over bedrock, then pass under the bridge. The bucket of flowers so common to the area stands out front.

The Vermont Division of Historic Sites described the surroundings when nominating the bridge for the National Register of Historic Places: "The bridge stands next to the 19th century mill district of the village. Together with an upstream dam and pond, the covered bridge and the mill buildings constitute an exceptionally attractive and functional historic element."

Photographic evidence from the 1870s indicates that an open bridge served at this location before it was replaced by the present covered bridge. The house at the west approach has changed little in outward appearance since then. The brick building beside the bridge was a blacksmith shop, first established in 1791. The first bridge at the site was built in 1797 and destroyed by flood the following year. The replacement bridge served until 1815 and was followed by a third. The dam above the bridge ran a gristmill.

The Agency of Transportation recommends that the Mill Bridge be rehabilitated for moderate traffic by replacing the current floor system and increasing the posted tonnage.

Larkin Bridge

The Larkin Bridge stands on Larkin Road east of Route 110, approximately 2.6 miles north of Tunbridge Village. There is a name board on each gable proclaiming the

The Larkin Bridge stands over a ford in the First Branch. Notice that because the bridge does not cross the stream squarely, the trusses are skewed.

name and a build date of 1902. The Larkin is peculiar in that the two trusses, while parallel, are noticeably skewed to fit the abutments where the bridge does not cross the stream squarely. The sixty-eight-foot multiple-kingpost structure, built by Arthur Adams, appears to be in good condition. There is a fording place nearby that shows continuing use. The Larkin Bridge is also decorated with potted flowers during the summer.

In the interest of preserving the historic bridge, the Agency of Transportation recommended that the structure be closed and bypassed or undergo rehabilitation to support moderate traffic.

Larkin Bridge

TOWN: Tunbridge
DATE: 1902
TRUSS LENGTH: 67'8"
BUILDER: Arthur Adams
TRUSS TYPE: Multiple kingpost

Flint Bridge

The Flint Bridge serves Bicknell Hill Road in a valley at the base of a conical hill. The eighty-eight-foot structure is a massive variety of queenpost believed to have been built in 1845. Gilbert Newbury finds the bridge remarkable for what he terms "a very interesting tension splice in the bottom chords."

Bridge restorer Milton Graton worked on the bridge in 1969, rebuilding the ends, replacing the camber and

> **Flint Bridge**
>
> TOWN: Tunbridge
> DATE: 1845
> TRUSS LENGTH: 87'7"
> BUILDER: Unknown
> TRUSS TYPE: Queenpost

the floor system, and installing new siding. He also raised the bridge for better drainage and capped the abutments. The state bridge inspector described the work as an "extensive restoration sensitive to the bridge's original fabric [and the work on the] bridge is an outstanding example of functional preservation of an historic structure."

Bicknell Hill Road leaves Route 110 to the east 1.8 miles north of the Larkin Bridge, or 0.2 miles south of the Tunbridge-Chelsea town line marker. The turn is currently marked only by a sign saying "Justin Morgan Memorial." A low and rambling brick dwelling stands at the intersection.

The Agency of Transportation recommends closing the structure and constructing a bypass as the best long-term action to preserve the bridge for the future. "[Because the] bridge is used by a school bus, emergency services, fuel trucks and snow plows, carrying capacity is an issue ... [and] trailer trucks can't use the bridge."

CHELSEA

The township of Chelsea was granted by the Province of New York about the year 1770 under the name of Gageborough. In November 1780 the town was granted to

The Flint Bridge is noted for the massive queenpost truss and the tension splices used in the bottom chords. The span was restored in 1969 by Milton Graton.

THE NORTHERN TRIBUTARIES OF THE WHITE RIVER

Bela Turner by the Vermont Legislature. It was originally chartered in August 1781 as Turnersburg.

Moxley Bridge

The Moxley Bridge, or Guy Bridge, is the northernmost of the string of bridges scattered along the First Branch of the White River. It stands on Moxley Road at a bend in the river next to a ford, and nicely scrolled signs at the gables declare "Moxley Bridge - 1883." The fifty-four-foot queenpost structure is attributed to builder Arthur Adams and the Vermont Division of Historic Sites nomination for inclusion of the bridge on the National Register of Historic Places offers 1886 or 1887 for a build date.

Because the bridge crosses the stream at an angle, the structure is slightly skewed to meet the abutments. The finish includes pilaster moldings and, at the north end, a wooden ramp. Look for very long hand-hewn bottom chords. The structure has been maintained true to the original construction—the only modification has been the addition of distribution beams under the floor.

Moxley Road leaves the east side of Route 110, 0.7 miles north of the Flint Bridge or 2.3 miles south of the

> **Moxley Bridge**
>
> TOWN: Chelsea
> DATE: 1883
> TRUSS LENGTH: 53'10"
> BUILDER: Arthur Adams
> TRUSS TYPE: Queenpost

The Moxley Bridge is the northernmost in the string of bridges over the First Branch. It is noted for the long hand-hewn bottom chords. A wooden ramp at the north portal helps keep water away from the bridge deck.

junction of Route 113 in Chelsea. The road may be marked only by a weight limit sign. The bridge is in good repair and is in daily use.

Sources

Allen, Richard Sanders. *Covered Bridges of the Northeast.* Brattleboro, VT: Stephen Greene Press, 1957.

Congdon, Herbert Wheaton. *The Covered Bridge.* Middlebury, VT: Vermont Books, 1973.

Farnham, Euclid. *A Pictorial History of Tunbridge, Vermont.* Randolph, VT: The Herald Printery, 1981

Graton, Milton S. *The Last of the Covered Bridge Builders.* Plymouth, NH: Clifford-Nicol, Inc., 1978.

Hemenway, Abby Maria. *The Vermont Historical Gazetteer, Vol. II.* Burlington, VT: A. M. Hemenway, 1868.

Herwig, Miriam. *Randolph's Beginnings.* Randolph Center, VT: A Greenhills Book, 1981.

Vermont Agency of Transportation Covered Bridge Study. Prepared for the State of Vermont Agency of Transportation by McFarland-Johnson, Inc., Binghamton, New York, 1995.

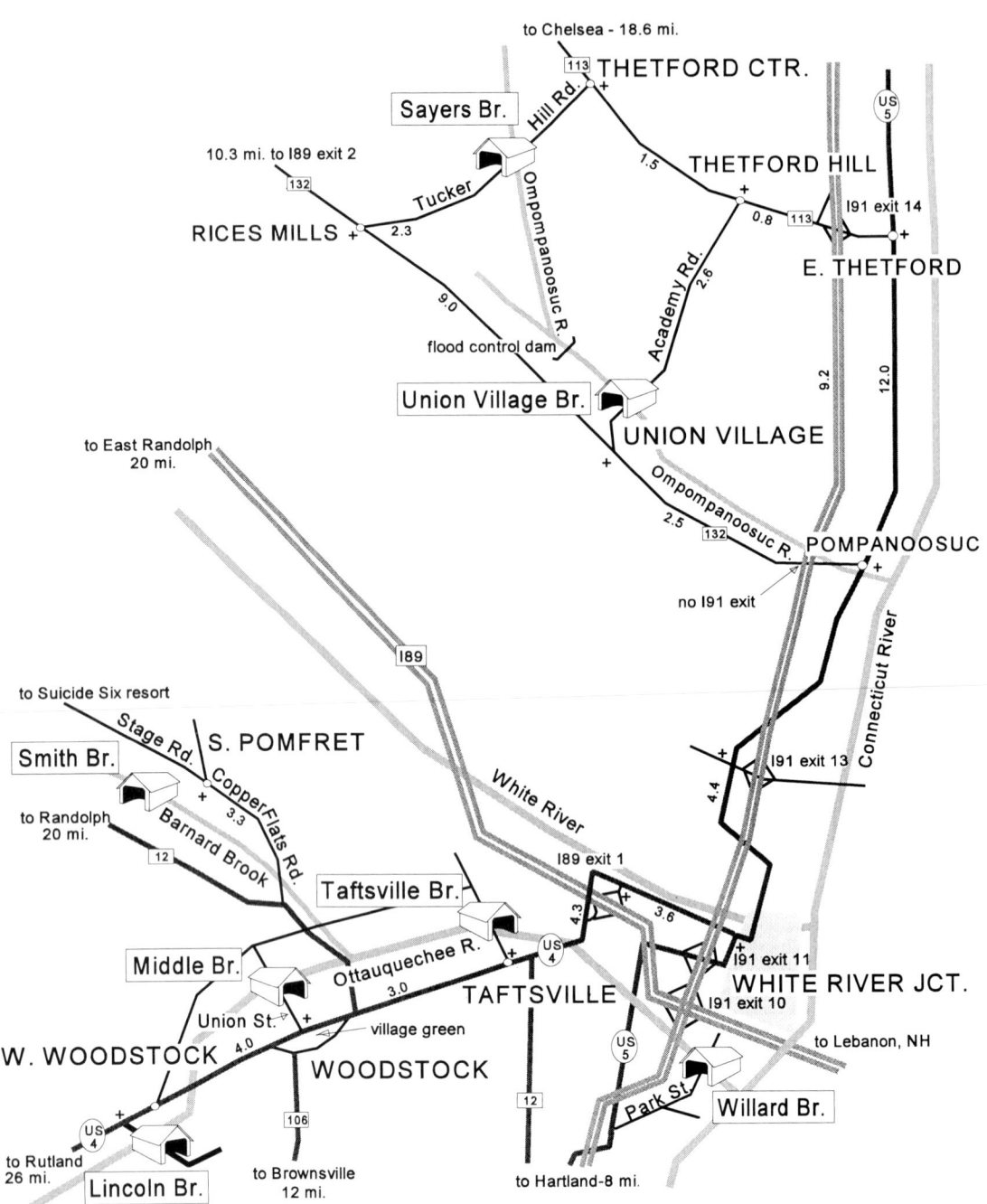

Tour 12
+ = Distance marker
Distances in miles (estimated)
Map not to scale
© 1996 Joseph C. Nelson

WOODSTOCK
Tour 12

Interstate 91 is handy for getting to the seven wooden bridges scattered from rural Thetford and the old mill town of North Hartland to Woodstock, a pretty, upscale resort town. I-91 is quick, but old Route 5 is more fun—it threads through the countryside, bringing the visitor to a world of small villages.

THETFORD

Thetford Town was chartered in 1761 and settled by people from both New Hampshire and Connecticut. In the 1860s historian Isaac Hasford remarked that while the Connecticut River Valley town was above average in thrift and population, nothing had happened there to claim space in history. "We have, beside farming, a riotous mill stream, the Oompompanusuc, bisecting the town, giving life and power to three smart villages. It has, on occasion, [washed] half their bridges and sometimes their mills down to the Connecticut and the town below."

Along the course of the "mill stream" were eight sawmills, four gristmills, a straw-board and paper mill, two flannel factories, a carriage shop and bedstead factory, and an edge tool and trip-hammer works.

When the Sayers Bridge is viewed from the top of the falls, the scene is hard to equal. The truss used in the span is a unique adaptation of Herman Haupt's 1839 patent. The midstream pier was added in 1963.

Sayers Bridge

The Sayers Bridge, thought by some to be a Haupt Truss[1] span, is the only one of its kind in New England, and one of just three in the United States. While the names of the builders are lost, the truss designer is remembered by Civil War buffs as the colonel who built and ran the U.S. Military Railroad in the South for Union forces.

Herman Haupt graduated from West Point in 1835. He resigned his commission to become district superintendent and chief engineer for the Pennsylvania Railroad. When the war began he was drafted to serve as superintendent of military railroads. He pushed his tracks through Virginia, building trestles out of found materials described by Abraham Lincoln as "bean poles and corn stalks."

Sayers Bridge

TOWN: Thetford
DATE: Unknown
TRUSS LENGTH: 129'
BUILDER: Unknown
TRUSS TYPE: Haupt with arch

---

[1] Jan Lewandoski: "Some people think that the Sayers Bridge is a Haupt Truss, mostly because some of the diagonals cross more than one bridge panel, but there is no evidence that the builder knew he was building a Haupt Truss." (See the section on Trusses.)

The Haupt design, as used here, resembles a multiple-kingpost truss. It differs in that it is assembled from planks instead of square timbers and is joined with treenails rather than with mortise and tenon. The builder integrated the whole with a segmented plank arch.

The bridge was strengthened in 1963. The existing roadway was replaced with a nail-laminated timber deck on four steel beams supported in mid-span by a concrete pier.

The Sayers Bridge is easily reached from I-91 Exit 14. Drive west two miles on Route 113, then take Tucker Hill Road south through Thetford Center. Sayers Bridge crosses the Ompompanoosuc River above a millpond. The river flows over a ruined dam and cascades down terraces of bedrock.

Union Village Bridge

The Union Village Bridge is in the center of the village and has spanned the river here since 1867. With a truss length of 113 feet, it is the longest multiple-kingpost span in the state—the average truss length of multiple-kingpost bridges in Vermont is fifty-four feet. An attempt was made

The Union Village Bridge, the state's longest multiple-kingpost span, has been defying gravity for more than 130 years.

to stiffen the structure with what has been called a kingpost arch. The Agency of Transportation bridge inspectors found that the long inverted "V" bracing has been lending little structural support "due to lack of substantial connection to the trusses."

The timber deck was replaced and the unmortared stone abutments capped in the 1970s, and the east abutment was faced with concrete. Except for the addition of distribution beams tie-bolted under the deck beams and the "kingpost" arches, the span remains as it was originally designed.

Unfortunately, the bridge is in trouble. The camber has reversed, so that, instead of arching upward at midspan, the old bridge is sagging. The Agency of Transportation has recommended that the Union Village Bridge be closed to traffic and bypassed or rehabilitated with a self-supporting roadway.

Find Union Village by returning to Route 113 from Thetford Center and driving east to Thetford Hill and Academy Road. Go south 2.5 miles to the bridge. Union Village Bridge stands in the Ompompanoosuc River valley in the shadow of the Union Village Flood Control dam.

> **Union Village Bridge**
>
> TOWN: Thetford
> DATE: 1867
> TRUSS LENGTH: 112'8"
> BUILDER: Unknown
> TRUSS TYPE: Multiple kingpost

NORTH HARTLAND

Hartland, then called Herford, was chartered by Benning Wentworth in July 1761. Oliver Willard was the first settler in town, arriving as early as 1762. In 1872, 110 years later, Oliver Brothers purchased the land at the falls from P. K. Willard to build the Ottauquechee Woolen Mill. The mill was served by a causeway with two bridges. The two bridges survived the 1927 flood, but the mill building was wrecked. One of the bridges, a queenpost span, was lost in the hurricane of 1938. The surviving bridge, the Willard, is named for the original land-owning family.

Willard Bridge

The Willard Bridge crosses the Ottauquechee River by North Hartland above a dam built on the crest of a natural waterfall. The mill pond above the dam, once used to run a woolen mill, is now a reflecting pool beneath the bridge. The stone foundations of the mill can be seen

This view of the Willard Bridge was taken from the bed of the Ottauquechee River below the woolens factory dam. The wooden cribbing used to retain the mill pond can be seen atop the bedrock escarpment.

Willard Bridge

TOWN: North Hartland
DATE: c. 1919
TRUSS LENGTH: 123'5"
BUILDER: Unknown
TRUSS TYPE: Plank-lattice

below the road at the east end of the bridge. A dirt road, also at the east end, leads to a park and easy access to the river below the dam.

The 123-foot plank-lattice span is in very good condition. The sides are board and batten, part of a renovation done in 1953. The Willard Bridge has been placed on the National Register of Historic Places by the U.S. Department of the Interior.

The quickest way to North Hartland and the Willard Bridge from Union Village is to retrace Route 113 back to I-91 Exit 14. Drive south to Exit 11 and Route 5. Alternatively, drive through the Union Village Bridge to Route 132, and then east to Pompanoosuc. From there, take Route 5 south through White River Junction. Look for Park Street, east off Route 5, about four miles south of I-91 Exit 11, just below the I-91 overpass in North Hartland. The bridge is located one-half mile east on Park Street.

Woodstock

In 1761 Benning Wentworth, royal governor of New Hampshire, granted the township of Woodstock to David Page and others. Settlement of Woodstock and many other towns in the area was delayed, however, because of the confusion caused by the king's order in council of July 1764 that made the western shore of the Connecticut River the eastern boundary of New York. The wording of the edict was such that those with New Hampshire titles deemed them still good, while others saw the territory open for land grants by New York.

Wentworth named Oliver Willard, a lawyer from Ashburnham, Massachusetts, to act as moderator for the Woodstock's first town meeting. By 1768 Willard had accumulated a lot of land in the town by buying out nervous claimants for low prices. He presented a petition for a royal charter to New York's Governor Tryon and received a grant in February 1772. Willard became a partisan of New York and eventually settled in the neighboring town of Hartland. It was his family that gave the Willard Bridge its name.

The Woodstock bridges—the Taftsville Bridge, the Middle Bridge, and the Lincoln Bridge—are all reached from Route 4. Route 4 is accessed by way of I-91 Exit 11, or from Route 5. The Smith Bridge stands north of Woodstock in the Town of Pomfret.

Taftsville

Taftsville's first settler, Stephen Taft of Mendon, Massachusetts, arrived in the early 1790s. He built a shop on the south side of the Ottauquechee River, where he made axes, scythes, and other edged tools. In 1794 he built a dam and a sawmill.

The Turkey Bridge

The first bridge in Taftsville was built in 1793 at the foot of Main Street, upstream of the present bridge. It was a shaky affair, and the townspeople feared it would crash down under them. The bridge had a peculiar truss featuring posts set in pairs joined at the top with a cross beam over the roadway. The cross beams were put to use as a roosting place for market-bound turkeys, but it is not known what other purpose they might have served.

The Taftsville Bridge uses a very complicated truss designed by a sophisticated framer. Spanning a crag below the power house dam, it may attract more painters than any other covered bridge in Vermont.

Taftsville Bridge

TOWN: Woodstock
DATE: 1836
TRUSS LENGTH: 189'2"
BUILDER: Solomon Emmons
TRUSS TYPE: Unique

Taftsville Bridge

Painted barn red, the Taftsville Bridge stands over the Ottauquechee River below the powerhouse in the village of Taftsville. The approach from the north side of Route 4 is steep and narrow.

The 189-foot structure rests on a center pier set in the gorge below the powerhouse dam. Auto, bicycle, and foot traffic on the single-lane span is heavy. North of the far portal, there is a wide spot in the road where artists often work. The Taftsville Bridge is probably one of the most painted bridges in New England.

The bridge was built by Solomon Emmons III and his son in 1836, and the Emmonses maintained the bridge afterward. The son, Edwin, did major repair work after the flood of 1869. In the bridge's early years, the funds for its maintenance came from the neighboring towns of Pomfret, Hartland, Woodstock, and Hartford. After 1851, when part of Hartland was ceded to Woodstock, Woodstock assumed ownership and maintenance of the bridge.

The truss is variously described as nondescript or mongrel. Jan Lewandoski, who inspected the bridge for the Agency of Transportation Covered Bridge Report, found it to be "a very complicated truss, designed, I think, not by a regular bridge builder, but by a very sophisticated framer who built an almost uncategorizable truss. It doesn't share any features with any other bridge. In general, it has queenpost and multiple kingpost elements, and it has arches. The posts are chestnut. It has gigantic bottom chords—seventy-foot eighteen-by-sixteen-inch timbers spliced in the middle. The splice is twenty feet long."

The four laminated arches were added to the bridge in the early 1900s. The Taftsville is one of the few arch bridges in Vermont where the arch extends below the bottom chords to the abutments, and, according to Lewandoski, the arch system is really working.

Because the two clear spans differ in length, the two sets of arches are not uniformly configured. The south span measures eighty-nine feet from abutment to pier; the other span to the north abutment is one hundred feet long. The long pair of arches consists of twelve laminated planks, and the short pair, ten.

The span was renovated in 1953. The bridge was raised on its abutments and pier to prevent water from flooding onto the bridge deck from the approach roads, and steel gussets were added in place of knee braces overhead. Except for the distribution beams installed under the deck system in the late 1980s and the steel gussets, the bridge remains as the Emmonses left it.

Middle Bridge

The Middle Bridge, or Union Street Bridge, stands in the heart of Woodstock, almost lost in the clutter of historic old facades. The 125-foot span over the Ottauquechee was built in 1969, replacing an 1877 iron bridge condemned in 1966.

According to the Woodstock Historical Society, it is the first authentic Town-lattice truss highway covered bridge to be built in Vermont since 1895. The truss is based on Ithiel Town's patent dated 1820.

The bridge was built by Milton Graton and Sons of Ashland, New Hampshire. The lattice planks are Oregon Douglas fir, and the treenails pinning them together are New Hampshire white oak. The State of Vermont paid

Middle Bridge
TOWN: Woodstock
DATE: 1969
TRUSS LENGTH: 125'
BUILDER: Milton Graton
TRUSS TYPE: Plank-lattice

Woodstock's Middle Bridge is the first public covered bridge built by the State of Vermont since 1889. Builder Milton Graton constructed his masterpiece near the crossing, then had it pulled across the river by Ben and Jo, a team of oxen, in July 1969.

50 percent of the cost, with the balance raised by private subscription.

The new bridge was set afire by juvenile pranksters on May 11, 1974, the night of the fireman's ball. The bridge structure was saved by the heroic efforts of the volunteer fire department and the fireproofing system installed by Graton. Graton completed restoration of the bridge in 1976, replacing the roof and siding. Char on the truss members was removed by sandblasting.

Lincoln Bridge

The Lincoln Bridge, named for a family who owned land nearby, crosses the Ottauquechee west of Woodstock. The portals stand within a few feet of the south median of Route 4, serving Fletcher Hill Road.

The 136-foot span employs a truss implemented in wood nowhere else in the U.S. According to town records, R. W. and B. H. Pinney built the bridge in 1865. Richard Sanders Allen, however, offers a build date of 1877 in his *Rare Old Bridges of Windsor County*. The truss used is similar to that patented by Thomas and Caleb Pratt in

The Lincoln Bridge uses what is thought to be a rare example of the Pratt truss constructed in wood. The bridge deck is suspended from a laminated wooden arch by crossed steel rods.

1844. The patent describes parallel chords used in conjunction with a boxed truss featuring wood vertical members with iron rod cross-braces supporting the lower chords. In the Lincoln Bridge, the upper chords are shallow arches. The boxed truss was never popular in the wood and iron implementation, but when iron and steel began to be widely used, it became the basic truss for bridge building. Look for it, with its boxed x's, in modern construction.

The bridge was renovated in 1947 and the Wright Construction Company renovated it again in 1989. Except for the high strength steel rods added below the bottom chords to add tensile capacity and distribution beams installed under the floor, the span remains true to the original builder's intent.

Thomas was the same Pratt who improved the Town-lattice truss for railroad bridges. The Town-Pratt truss is used in the Fisher Bridge in Wolcott.

Lincoln Bridge

TOWN: Woodstock
DATE: 1865
TRUSS LENGTH: 135'10"
BUILDER: R. W. & B. H. Pinney
TRUSS TYPE: Pratt arch

Smith Bridge at Pomfret

The Smith Bridge, privately owned, stands on a portion of the Lawrence Rockefeller holdings in Pomfret. It

> **Smith Bridge**
>
> TOWN: Pomfret
> DATE: 1973
> TRUSS LENGTH: 40'
> BUILDER: Cummings Construction
> TRUSS TYPE: Plank-lattice

The plank-lattice truss in the Smith Bridge at Pomfret was taken from a retired bridge in Garfield, Vermont. Placed here as a gateway to a housing development, the bridge has since blended into the community.

can be seen just north of the village of South Pomfret, where it crosses Barnard Brook in an open field. Jim Mclaughlin, who lives next to the bridge access path, is generally considered to be the caretaker, an office ascribed to him only by his proximity. Actually, the whole neighborhood looks after the bridge, which is watched zealously to see that it comes to no harm. It is suggested that viewers approach the bridge with care and not stray off the pathway, especially if the field has been planted with crops.

According to Mclaughlin, the bridge is used almost exclusively by the farmer who leases the fields to move his equipment across the brook. Other than that, parties are occasionally held there in good weather—in fact a recent June wedding was planned, but it was rained out by an exceptionally wet early summer day.

The forty-foot span was assembled here in 1973 by the Cummings Construction Company from plank-lattice trusses salvaged from the one-hundred-foot Garfield Bridge. The Garfield Bridge, built in the 1870s, crossed the Green River in Garfield Village in Hyde Park. When work began in 1946 on the Green River Reservoir project

upstream, the old bridge was strengthened to handle construction traffic. It was abandoned in 1965, when the town bypassed it with a culvert. J. P. Rich, president of a local surveying firm, purchased it in 1971 to ensure it would be preserved.

Thurston Twigg-Smith, Jr., of ASA Properties Vermont, Inc., a real estate development company, bought the Garfield Bridge to provide access to two of the corporation's properties, one in Pomfret, the other in West Windsor. The trusses were taken down, cut in half, and trucked to the building sites. The two bridges were never formally named, but the developer referred to them as the Pomfret Bridge and the Ascutney Bridge.

Lacking bedrock, the abutments of the Pomfret bridge were built on six wooden pilings driven thirty-two feet into the stream bank. Construction was completed on the forty-foot bridge in 1973. It is handsome, with extended gable-ends and with sides left open to display the lattice truss. The roof is finished with shingles imported from Australia.

ASA Properties did not complete its development project in Pomfret. The subdivision was challenged by strict zoning codes and a petition against the building of the bridge. The opposition was not to the bridge itself, it was found, but to the housing tract to which it would provide access. The land was bought by the Suicide Six Ski Resort's parent company.

The Smith Bridge in Pomfret is easily reached from Woodstock by taking Route 12 where it leaves Route 4 north from the center of the village. Just follow the signs to Suicide Six Ski Resort—drive approximately one mile to Copper Flats Road where it forks to the right, then 1.7 miles to South Pomfret. Turn left on Stage Road toward the ski resort and watch to the left for the bridge.

Sources

Allen, Richard Sanders. *Covered Bridges of the Northeast.* Brattleboro, VT: Stephen Greene Press, 1957.

Allen, Richard Sanders. *Rare Old Covered Bridges of Windsor County.* Brattleboro, VT: Stephen Greene Press, 1962.

Dana, Henry Swan. *The History of Woodstock, 1761-1886.* Woodstock, VT: Countryman Press, 1980.

Graton, Milton S. *The Last of the Covered Bridge Builders.* Plymouth, NH: Clifford-Nicol, Inc., 1978.

Hagerman, Robert L. *Garfield Bridge.* Unpublished manuscript.

Hemenway, Abby Maria. *The Vermont Historical Gazetteer, Vol. III.* Burlington, VT: A. M. Hemenway, 1877.

In Sight of Ye Great River. Hartland, VT: Hartland Historical Society, 1991.

Jennison, Peter S. T*he History of Woodstock, 1890-1983.* Woodstock, VT: Countryman Press, 1983.

Vermont Agency of Transportation Covered Bridge Study. Prepared for the State of Vermont Agency of Transportation by McFarland-Johnson, Inc., Binghamton, New York, 1995.

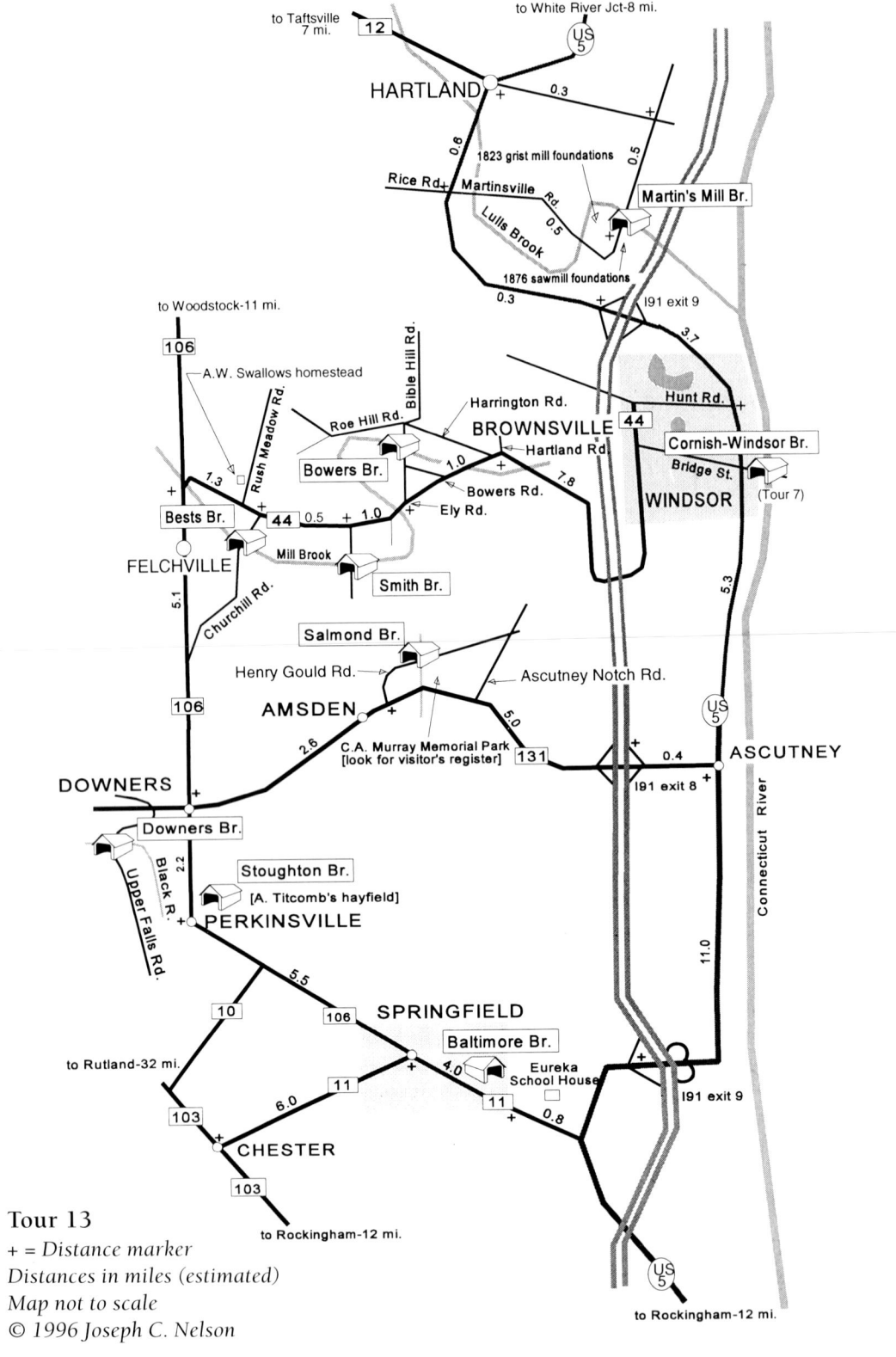

Tour 13
+ = Distance marker
Distances in miles (estimated)
Map not to scale
© 1996 Joseph C. Nelson

The WINDSOR AREA

Tour 13

There are eight covered bridges in the countryside between Hartland and Springfield. Three of them occupy special places in the communities of Weathersfield and Springfield because the people there made special efforts to keep them. Two were rescued from a dam project, and another was saved from demolition to become a monument to a prominent citizen. All of the bridges are historical treasures, and some of them are especially noteworthy. Three are attributed to master builder James Tasker, and two others are rare tied-arch structures.

All of the sites are easily reached from U.S. Route 5 and I-91.

Martin's Mill Bridge

James Tasker, famed builder of the Cornish-Windsor Bridge, put a 137-foot plank-lattice bridge across Lulls Brook next to Martin's Mill in Hartland in 1881. While Martin's Mill has become broken concrete and scattered stone, Martin's Mill Bridge still serves Martinsville Road.

The water power provided by Lulls Brook gave rise to a small industrial center that is now gone. The ruined buildings by the south portal of the covered bridge and the foundations along the south bank of the brook are the remains of Martin's sawmill and lumber business. A wooden penstock can still be seen standing among the

---

Martin's Mill Bridge

TOWN: Hartland
DATE: 1881
TRUSS LENGTH: 136'11"
BUILDER: James Tasker
TRUSS TYPE: Plank-lattice

Martin's Mill Bridge stands in the midst of the remains of a nineteenth century industrial complex. The bridge is said to have once carried a penstock to power the works here.

trees along Route 12. Richard Sanders Allen, in his *Rare Old Bridges of Windsor County*, tells us that the bridge once supported a wooden "conduit" that carried water to the mill.

In 1979 distribution beams were tie-bolted under the deck, diagonal steel sway braces were installed, and steel cables were added to lend lateral support. Except for additions like these, James Tasker's bridge remains much as he left it more than one hundred years ago.

Martinsville Road is clearly marked at the junction with U.S. Route 5. Travelers from the north on U.S. Route 5 will find Martinsville Road about one-half mile south of the junction with Route 12. Turn left to go to the bridge. Travelers on Route 5 and I-91 can proceed north from the I-91 Exit 9 clover about 0.3 miles to turn right on Martinsville Road.

Bowers Bridge

The Bowers Bridge, or Brownsville Bridge, spans Mill Brook amid open fields and low rolling hills, serving the road to the hamlet of Sheddsville.

Built in the early 1900s by an unknown craftsman, the forty-five-foot span uses a simple arch-truss con-

> **Bowers Bridge**
>
> Town: West Windsor
> Date: c. 1919
> Truss length: 45'4"
> Builder: Unknown
> Truss type: Tied arch

structed of a laminate of five ten-inch planks. The chords are suspended from the arches on three-quarter-inch iron rods, the whole protected by a post-and-beam shed set upon the bridge deck. The abutments are unmortared stone slabs with concrete caps. Bowers Bridge is essentially a copy of the older Bests Bridge two miles to the west.

Bowers and Bests bridges are two of the last three surviving tied-arch bridges in the state—the third is the Lakeshore Bridge in Charlotte. The originator of the design of this unique truss is unknown.

Bowers Bridge is found north of Route 44 on Bible Hill Road, six miles west of Windsor and one mile west of Hartland Road in the village of Brownsville.

Smith Bridge at Brownsville

The Smith Bridge stands over Mill Brook in the valley below Mount Ascutney. Privately owned, the short plank-lattice bridge is part of a project undertaken by Thurston Twigg-Smith, Jr., of ASA Properties Vermont, Inc., a real estate development company. He bought a bridge that once served Garfield, Vermont, to provide access to two of the corporation's properties, this one in West Windsor,

Bowers Bridge uses a rarely seen tied arch truss. The span is exceptionally trim in appearance. The truss is protected from the elements by a post and beam shed placed on the bridge deck.

The Smith Bridge in West Windsor is a twin to the one in Pomfret, built from the trusses of the retired Garfield Bridge. Here, the internal bracing was reworked to admit the passage of construction trucks.

the other in Pomfret. The old bridge was disassembled and the trusses, one hundred feet long, were cut in half and trucked to the two building sites. The developer referred to this bridge as the Ascutney Bridge, but it was never formally named.

Both of Twigg-Smith's bridges were assembled in 1973 by the H. P. Cummings Construction Company. The two bridges are similar in appearance, handsome with extended gable-ends and with sides left open to display the lattice truss. The roof shingles, unique in appearance, were imported from Australia. The "Ascutney" Bridge has since been altered to increase passage height—large rectangular pieces of the portals have been cut out and the upper bracing system changed. A sign on the gable-end declares: "13 ft 40 ton." It rests on cast concrete abutments with two large steel beams supporting the timber deck.

The Smith Bridge crosses Mill Brook south of Route 44 and two miles west of Brownsville. For more on the Garfield Bridge story, refer to the section on the Smith Bridge at Pomfret in Chapter 12 (Woodstock).

Smith Bridge at Brownsville

TOWN: West Windsor
DATE: 1973
TRUSS LENGTH: 39'10"
BUILDER: Cummings Construction
TRUSS TYPE: Plank-lattice

Bests Bridge

Bests Bridge stands hidden in a cluster of buildings south of Route 44 on Churchill Road, 2.5 miles west of Brownsville, or 1.3 miles east of the junction with Route 106 north of Felchville. Bests Bridge, also known as Swallows Bridge, was built over Mill Brook by A.W. Swallows in 1890. It is structurally a smaller twin to Bowers Bridge.

The bridge was refurbished in 1991. The Agency of Transportation Covered Bridge Inspection report comments: "Technically unsophisticated, it serves adequately the need for an economically durable span to carry a lightly travelled country road over a brook."

Bests Bridge

Town: West Windsor
Date: 1890
Truss length: 37'5"
Builder: A. W. Swallows
Truss type: Tied arch

Bests Bridge, a twin to the Bowers Bridge, may have been built thirty years earlier. The arches were originally constructed of a laminate of five planks—two more planks were recently added to the bows.

WEATHERSFIELD

Weathersfield was a Benning Wentworth grant dating back to Aug. 20, 1761. The town was named for Weathersfield, Connecticut (now Wethersfield).

Weathersfield is home to three unique bridges. One of them, Downers Bridge, is an example of beauty in bridge architecture. The others are monuments to the

dedication local residents showed in preserving something of the old ways in the face of inexorable change.

In the 1950s the U.S. Corps of Engineers conceived a flood control project for the Springfield-Weathersfield area. The plan called for a dam in the Black River valley to collect excess runoff and divert it harmlessly downstream to the Connecticut River. Naturally, anything above the dam would be subjected to periodic immersion, so the policy was to clear everything out of the collection basin—all the houses and bridges had to go! The Salmond and Stoughton Bridges stood over the Black River near Stoughton Pond, part of the flood control basin.

The Weathersfield Historical Society came to the rescue, alerting the townspeople to the threat to their heritage. In response to community concerns, the flood control project contractor hired Milton Graton and Sons of Ashland, New Hampshire, to move the bridges. In 1959 the Salmond Bridge was relocated in the Amsden Village road maintenance yard. Graton and Sons moved the Stoughton Bridge to Perkinsville.

Salmond Bridge

The Salmond Bridge stands east of Amsden Village, on Henry Gould Road. Henry Gould Road leaves Route 131 to the north, five miles west of the junction with U.S. Route 5. The townspeople joined to rescue their bridge from the town equipment yard and relocate it over a small stream beside a little park with picnic tables. There are two bronze plaques placed there by the Weathersfield Historic Society, one for the bridge and one for the park. The bridge plaque says:

Salmond Bridge

TOWN: Weathersfield
DATE: 1880
TRUSS LENGTH: 53'2"
BUILDER: James Tasker
TRUSS TYPE: Multiple kingpost

> **SALMOND BRIDGE**
> Salmond Bridge was built by James H. Tasker circa 1880. This 53' Multiple Kingpost structure spanned the Black River near Stoughton Pond and was named after the Salmond family living near the bridge. It remained in this area until 1959 when it was relocated beside Route 131 in Amsden in order to remove it from the flood control area. There it was used as a town storage shed. It was restored and moved to this site over Sherman Brook in 1986 through the efforts of the townspeople of Weathersfield.

James Tasker's Salmond Bridge has caring friends. It came to this place beside a park having twice been rescued from destruction.

Milton Graton and Sons presided over the second move of the Salmond Bridge and did the restoration work.

Downers Bridge

The Downers Bridge is a truly beautiful bridge with classic lines. Built around 1840, the portals are rendered in the Greek Revival mode. Partial cornice returns and enclosed roof-end overhangs bracket the splined gable-end boarding. The treatment simulates the Greek pediments in the Federal architecture popular in the first half of the 1800s. The Sanderson Bridge over Otter Creek in Brandon, built circa 1838, is the only other covered bridge in Vermont with cornice returns.

Downers Bridge is Vermont's last example of splined boarding in a bridge portal. The 121-foot plank-lattice span, also known as the Upper Falls Bridge, crosses high over the Black River. The tall stone abutments and long curved wingwalls are the most impressive examples of dry stone masonry in the state. The northern abutment has not been cased in concrete and remains a tribute to the skill of the masons who provided the foundations for Vermont's covered bridges.

Downers Bridge

TOWN: Weathersfield
DATE: c. 1840
TRUSS LENGTH: 121'5"
BUILDER: James Tasker
TRUSS TYPE: Plank-lattice

Downers Bridge stands high over the Black River. It is worthwhile to take a walk down to the river's edge, where the abutments can be viewed. Notice the mill foundations along the river bank.

The trusses were originally constructed with sway braces, or buttresses, but they are no longer in evidence. They may have been dispensed with when the restoration was done by Milton Graton and Sons in 1975.

The south bank of the river, upstream of the bridge, is lined with the foundations of abandoned mill works. Downers Bridge is found south of Route 131 on Upper Falls Road, 0.2 miles west of the junction with Route 106.

Stoughton Bridge

When the Stoughton Bridge was rescued from the flood control area, the little bridge was given refuge from a changing world on Andrew Titcomb's farm in Perkinsville. The homeless bridge was set over a ditch at the edge of a hay field east of Route 106 just south of Perkinsville.

The Stoughton Bridge, which is now often called the Titcomb Bridge, was built in 1880 by James and Henry

---

Stoughton Bridge

TOWN: Weathersfield
DATE: 1880
TRUSS LENGTH: 46'2"
BUILDER: James Tasker
TRUSS TYPE: Multiple kingpost

Andrew Titcomb found a place in the back of his meadow for the Stoughton Bridge when it was saved from the Weathersfield-Springfield flood control project. Now often called the Titcomb Bridge, it looks right at home fronting the farm woodlot.

Tasker using a multiple-kingpost truss. The Gratons set the bridge on abutments constructed of stone from the original site. Andrew Titcomb, an architect, did the restoration work.

Baltimore Bridge

The Baltimore Bridge and the Eureka School are located in a small park next to Route 11 east of Springfield, less than one mile from U.S. Route 5 and I-91 Exit 9. For travelers driving south on Route 106, the site lies nine miles south of Perkinsville.

The school has a historic site marker, while a bronze plaque adorns the bridge.

> In memory of Senator Ralf E. Flanders, 1880-1970. This last covered bridge in Springfield Vermont, the Baltimore Bridge, completed in 1870 was one of two built across Great Brook by Granville Leland and Denis Allen. It was moved from the original location on Baltimore Road, North Springfield and restored on this site by Milton S. Graton, covered bridge builder, Ashland, N.H. 1969-1970.

THE WINDSOR AREA

The old Baltimore Bridge is a fitting monument to Senator Ralf Flanders, who preserved both the bridge and the Eureka School House. The roof and moldings over the portals are evidence of tender loving care.

In 1967 the forty-five-foot plank-lattice truss bridge was declared unsafe. A committee was formed to move the bridge to the park beside the Eureka School house, and there restore it. The committee, headed by former U.S. Senator Ralf Flanders, contracted with Milton Graton. Graton moved the bridge seven miles from its original site over Great Brook, where it served Baltimore Village.

Baltimore Bridge

TOWN: Springfield
DATE: 1870
TRUSS LENGTH: 44'11"
BUILDER: Leland & Allen
TRUSS TYPE: Plank-lattice

Sources

Allen, Richard Sanders. *Covered Bridges of the Northeast.* Brattleboro, VT: Stephen Greene Press, 1957.

Congdon, Herbert Wheaton. *The Covered Bridge.* Middlebury, VT: Vermont Books, 1970.

Graton, Milton S. *The Last of the Covered Bridge Builders.* Plymouth, NH: Clifford-Nicol, Inc., 1978.

Hagerman, Robert L. *Garfield Bridge.* Unpublished manuscript.

Morse, Victor. *Windham County's Famous Bridges.* Brattleboro, VT: Stephen Greene Press, 1960.

Vermont Agency of Transportation Covered Bridge Study. Prepared for the State of Vermont Agency of Transportation by McFarland-Johnson, Inc., Binghamton, New York, 1995.

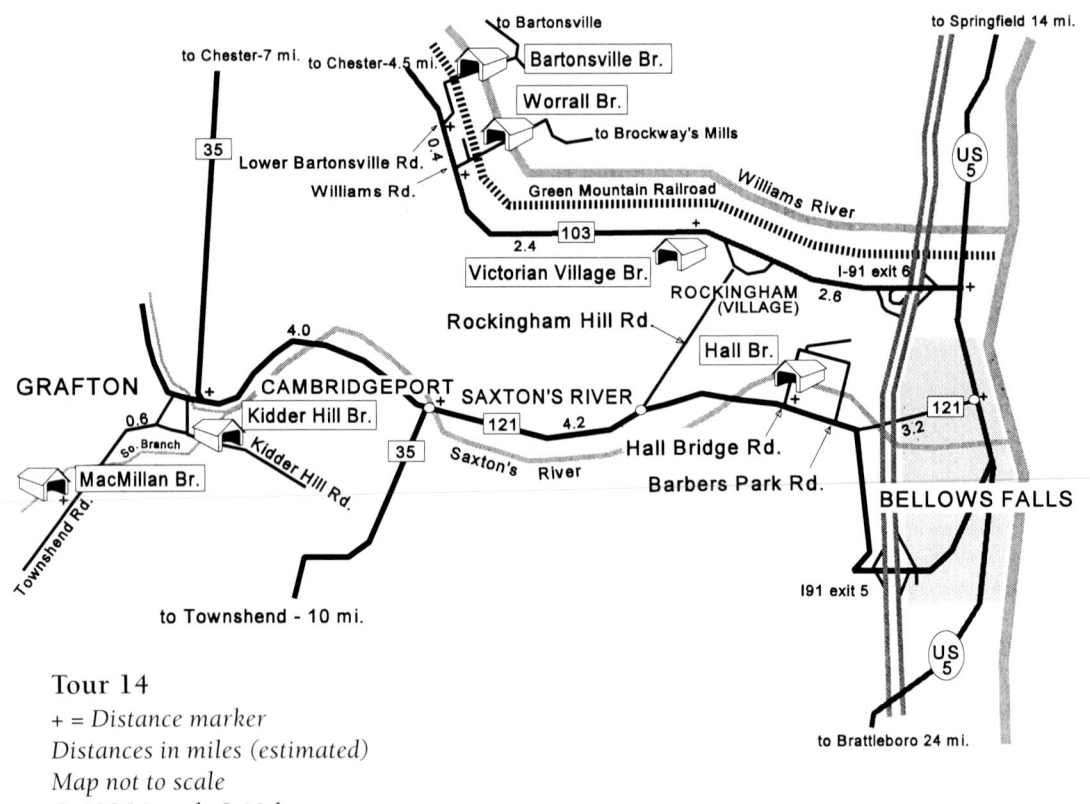

Tour 14
+ = Distance marker
Distances in miles (estimated)
Map not to scale
© 1996 Joseph C. Nelson

ROCKINGHAM to GRAFTON

Tour 14

All of the streams flowing from the east slope of Vermont's mountain ranges empty into the Connecticut River. In the southeastern part of the state, the major streams are the Williams River at Rockingham, Saxton's River at Bellows Falls, and the West River at Brattleboro. In the early days of the development of the region, these waterways provided access to a roadless interior and power to turn mill-wheels. People came, industry grew, roads were built, and bridges were thrown over the streams that nurtured it all.

The bridges can be easily reached by I-91 or by U.S. Route 5. This approach needs little description. For the more adventurous, the Scenic Route travels from bridge to bridge with no long-distance retracing of paths and provides an opportunity to see more of rural Vermont.

THE SCENIC ROUTE

For travelers from the north, the Scenic Route begins in Chester on Route 103. Drive south to find the Bartonsville Bridge at Bartonsville, then Worrall Bridge and the Victorian Village Bridge. Continue east to U.S. Route 5 and drive south to Bellows Falls. Here take Route 121 west to find the Hall Bridge. Continue west to Grafton and the Kidder Hill Bridge. Also in Grafton, find the MacMillan Bridge on Townshend Road, an attractive reproduction.

> **Williams River**
>
> The Williams River was named in honor of the Reverend John Williams, who preached the first sermon in Rockingham. After the French and Indians raided Deerfield, Massachusetts, in February 1704, they took their captives on a three-hundred-mile trek to Canada. They camped overnight at a place where a tributary flowed into the Connecticut River. The Reverend Williams was permitted to hold services there for his fellow captives on Sunday, March 5. He based the service on Lamentations 1:18.

The bridge seeker can retrace Route 121 back to Bellows Falls or turn right in Cambridgeport onto Route 35 south to Townshend to begin the tour of Vermont's old southeast towns.

ROCKINGHAM

The first English settlers in Rockingham were granted land by Massachusetts. First called Fallstown, then Great Falls, it became Bellows Falls when first proprietor Benjamin Bellows settled there.

Chartering the town in 1752, Benning Wentworth became interested in the early settlement of Rockingham when he heard reports of the timber growing along the River of Pines. He came in person to see that the huge pines were reserved for the use of the Royal Navy. Wentworth caused the first sawmills to be established, and one of them was located on the Williams River at a place that came to be called Brockway's Mills. Rockingham was named by Wentworth in honor of his cousin, Charles Watson Wentworth, Marquis of Rockingham.

According to historian Victor Morse in his *Windham County's Famous Bridges*, there were once seventeen covered highway and railroad bridges in Rockingham. Today, the town has four highway covered bridges. Three of them, all plank-lattice, were crafted by master builder Sanford Granger: Worrall Bridge, Bartonsville Bridge, and Hall Bridge. The fourth, the Victorian Village Bridge, was a queenpost structure that has been rebuilt as a kingpost bridge.

Bartonsville Bridge

Bartonsville Bridge

TOWN: Rockingham
DATE: 1870
TRUSS LENGTH: 150'10"
BUILDER: Sanford Granger
TRUSS TYPE: Plank-lattice

The Bartonsville Bridge opens at the edge of a railroad right-of-way. It can only be imagined what happened when an uninitiated horse came out of the bridge to meet its first steam locomotive.

The Bartonsville Bridge, also known as the Williams River Bridge, is shipshape and squared away for whatever weather and floods the fates may bring. In its suit of new siding and metal roof, it looks now much as it might have appeared in 1870 when Sanford Granger built it to replace the span lost in the flood of 1869. The gables are extended, keeping moisture away from the truss, and the portals are gently curved, a declaration of the builder's pride in his work.

The siding is ported on each end on both sides, exposing the plank-lattice truss. The ports let in air, light, and just enough of a view to give the traveler fair warning of an oncoming train—the 151-foot bridge ends within a few feet of the Green Mountain Railroad.

Except for steel rods installed under the roadway to increase lateral strength and the addition of distribution beams tie-bolted under the deck, the bridge structure remains essentially the same as it was when Mr. Granger completed it.

The bridge is found north off Route 103 on well-marked Lower Bartonsville Road.

Worrall Bridge

Worrall Bridge crosses Williams River on Williams Road. Williams Road is clearly marked and leaves Route 103 to the north. After crossing the bridge, the road continues on to Brockway's Mills Gorge. There, the Williams River flows between sheer hundred-foot cliffs after dropping down a series of cascades. Sanford Granger completed the eighty-three-foot bridge in 1868, just before the flood of 1869. The gorge must have been an awesome sight with the flood waters passing though it. The new bridge was very nearly lost.

Over the years, the Worrall Bridge has been modified, perhaps because it was built without the usual strengthening secondary chords. The lattice has been reinforced by six pairs of seven-by six-inch vertical posts and is further steadied by iron rods and steel cables. As with many other wooden bridges, distribution beams have been tie-bolted under the deck system, and the stone abutments have been encased in concrete. There is a twenty-foot timber ramp on steel beams at the northeast end that serves to drain excess road water away from the bridge deck.

> **Worrall Bridge**
> TOWN: Rockingham
> DATE: 1868
> TRUSS LENGTH: 82'8"
> BUILDER: Sanford Granger
> TRUSS TYPE: Plank-lattice

Worrall Bridge carries the flavor of old times with its weathered siding. Notice that, alone among Sanford Granger's plank-lattice projects, this span lacks secondary chords.

The Victorian Village Bridge, part of the Orton Country Store complex, is reminiscent of the simple kingpost spans that once numbered in the hundreds. Note that in addition to the usual king rods, other steel rods support the bridge deck from the middle of each kingpost brace.

Victorian Village Bridge

TOWN: Rockingham
DATE: 1967
TRUSS LENGTH: 43'9"
BUILDER: Aubrey Stratton
TRUSS TYPE: Kingpost

The exterior structure of the bridge is similar to that of the Bartonsville span, but the surroundings have a different ambience. Worrall Bridge stands next to a junk yard, which is surrounded by privacy fence. Bridge viewers should be aware that unfriendly junkyard dogs live behind the fence.

Victorian Village Bridge

The Victorian Village Bridge stands next to Orton's Vermont Country Store on the south side of Route 103 near the village of Rockingham.

The bridge was originally built in Townshend as the Townshend Depot Bridge by Harrison Chamberlin in 1872 using a queenpost truss. The bridge was taken down in 1959 by the U.S. Army Corps of Engineers to make way for the Townshend flood control project. The disassembled bridge was stored until 1967. It was then taken to the grounds of the new Country Store, shortened to just forty-four feet in length, and reassembled by Aubrey Stratton as a kingpost truss. The bridge now represents a class of small bridge that has largely disappeared over the years, many replaced by culverts. There are but four kingpost bridges still standing, in addition to the two spans that are part of the Scott Bridge.

Notice that iron rods serve in place of timber kingposts. Many of the later kingpost trusses were built this way, and some of the older bridges were repaired using augmenting iron rods, the rods being superior to wood in bearing tensile stress.

Hall Bridge

The Hall Bridge stands at the north side of Route 121, just 3.2 miles from U.S. Route 5. Sanford Granger, the premier builder in the area, built the original bridge in 1867 using the plank-lattice truss. Unfortunately, Granger's bridge collapsed under a truck in 1980. The 120-foot structure that now spans Saxton's River was erected in 1982 by Milton Graton and Sons as an authentic replacement, using traditional methods and materials.

A bridge historian commented that the span was originally constructed with what he described as "flying but-

The Hall Bridge is a fine reproduction of the span that served the locally famous amusement park here. Milton Graton's work is faithful even to the portal width of twelve feet, the narrowest in the state.

> ## Hall Bridge
>
> TOWN: Rockingham
> DATE: 1982
> TRUSS LENGTH: 120'1"
> BUILDER: Milton Graton
> TRUSS TYPE: Plank-lattice

tresses." These devices, no longer in evidence, were extended floor-beams with a diagonal brace to the truss used to supplement the lateral braces under the roof rafters. The purpose of the bracing is to keep the trusses standing upright and aligned against lateral forces from wind and water. Examples of buttresses in use can be viewed at the School House Bridge in North Troy and at the Scott Bridge in Townshend.

All of the bridges known to have been designed by Granger feature extended gables except this one. Here, only the gable-end roof overhang extends over the portals, and the gables are sheathed with clapboard rather than the usual vertical planking. The rounded portals are finished with "keystone" and pilaster molding.

The Hall Bridge has two unusual features. First, its portals are only twelve feet wide—true to the dimensions of the original—the narrowest portals of all of Vermont's covered bridges. Second, the planks used in the lattice and in the chords are four inches thick instead of the usual three inches. These planks are probably heavier than the originals. Cables have since been added to the exterior structure to give the bridge additional lateral support.

The original bridge was on the National Register of Historic Places. The replica has been designated a scenic resource by the town. It once served Barber Park, the town's celebrated amusement park and picnic grove. Today it serves unpaved Paradise Hill Road leading to Bellows Falls' village forest. The bridge has also been known as the Barber Park Bridge and the Osgood Bridge.

> ## The First Bridge Across the Connecticut
>
> The first bridge across the Connecticut River was built at Bellows Falls by Colonel Enoch Hale. Isaiah Thomas wrote in the *Massachusetts Spy*, February 10, 1785: "This bridge is thought to exceed any ever built in America in strength, beauty, and public utility." Hale's bridge was 365 feet long, supported in the middle by a pier built on a rocky island. The bridge was the first built in New England with a clear span longer than one timber. The famous Tucker Toll Bridge replaced Hale's bridge in 1840. The wooden Tucker Bridge was replaced by a concrete bridge in 1931.

Grafton

Grafton was chartered in 1754, but because of problems with the French and Indians, possession was not taken until 1763, and the first permanent settlement was not established until 1780. The township was originally named for John Thomlinson, Benning Wentworth's business representative in London. In 1791 the townspeople decided the town should not bear the name of an Englishman who had never seen the place. It was voted to auction the privilege of renaming the town and Joseph Axtell of Grafton, Massachusetts, won with a bid of five dollars and a jug of rum. Accordingly, the name of the town was changed from Tomlinson (the "h" had been dropped in 1788) to Grafton.

By 1824 the water power from six dams helped to support four sawmills, three gristmills, two fulling mills, two carding mills and two tanneries. There is also an extensive soapstone bed along the southern border of the town, and settlers used to chop slabs out of it for use as hearthstones. Commercial quarrying of the stone began in 1825.

Kidder Hill Bridge

The sixty-eight-foot Kidder Hill Bridge is the longest of Vermont's four surviving kingpost bridges. It dates from 1870, when it was built to replace a span lost in the flood of 1869. In 1938 the bridge survived what was termed a "500-year flood." It was left isolated but still standing with the road washed out at both approaches. Kidder Hill Road and the bridge once served a soapstone quarry, which is now closed. The road continues to be used to access the Bear Hills hiking and ski-touring area.

The bridge in its prime was deemed strong enough to support wagon-loads of stone, even though sixty-eight feet is quite long for a kingpost span—the three other surviving kingpost bridges average only forty-three feet in length. The kingpost braces measure eight inches square and the kingposts are iron rods one-and-three-quarter inches in diameter. The truss is augmented with a single buttress on each side, a unique feature for a non-plank-lattice bridge in Vermont.

Considering the bridge to be unsafe after some 120 years of service, the Agency of Transportation recon-

> **Kidder Hill Bridge**
> Town: Grafton
> Date: 1870
> Truss length: 67'7"
> Builder: Unknown
> Truss type: Kingpost

The Kidder Hill Bridge is the only kingposter in Vermont that still has sway braces. These braces were more commonly used on open bridges, which had no other provision for lateral bracing.

structed it in 1994–95 and made the roadway self-supporting. The method used is new to Vermont—a one-foot by five-foot by sixty-seven-foot laminated wood beam was installed on each side of the roadway inside of the covered bridge, the ends resting on the abutments. Wood floor beams measuring twelve-inches by fifteen-inches pass under the wooden deck, each beam end-bolted through the laminated beams above.

Grafton Village lies about twelve miles west of U.S. Route 5 on Vermont Route 121. The Kidder Hill Bridge crosses the South Branch of Saxton's River on Kidder Hill Road. The well-marked road leaves the south side of Route 121 at the west end of a concrete highway bridge.

The MacMillan Bridge is a beautifully executed replica constructed by a true lover of covered bridges. Its wooden truss serves no function other than to please the eye, however.

MacMillan Bridge

The MacMillan Bridge crosses the South Branch of Saxton's River a few feet south of Townshend Road, next to the Grafton Village Cheese Company.

A sign on the bridge proclaims: "Footpath to Windham Ponds." The very attractive sixty-foot bridge is actually a stringer bridge with the addition of a kingpost truss of convincing appearance. The MacMillan Bridge is a good reproduction, but its truss serves no function.

Iron rods are used as the tension member in the faux truss, rather than timber, imitating the nearby Kidder Hill Bridge and other authentic kingpost trusses. Notice that many of the timbers used in the interior here have mortises in them that serve no purpose, evidence that they were salvaged from a post-and-beam building.

The MacMillan Bridge is found alongside Townshend Road about one-half mile south of Route 121.

MacMillan Bridge

TOWN: Grafton
DATE: 1967
TRUSS LENGTH: 60'6"
BUILDER: S. MacMillan
TRUSS TYPE: Stringer (false truss)

SOURCES

Allen, Richard Sanders. *Covered Bridges of the Northeast.* Brattleboro, VT: Stephen Greene Press, 1957.

Congdon, Herbert Wheaton. *The Covered Bridge.* Middlebury, VT: Vermont Books, 1973.

Hemenway, Abby Maria. *The Vermont Historical Gazetteer, Vol. V.* Burlington, VT: A. M. Hemenway, 1891.

Hobson, Jane Baker. *Rockingham, Vermont Place Names - Why Is It Called That?* Bellows Falls, VT: The Model Press, no date.

Morse, Victor. *Windham County's Famous Bridges.* Brattleboro, VT: Stephen Greene Press, 1960.

Pettengill, Helen M. *History of Grafton, Vermont 1754-1985.* Grafton, VT: The Grafton Historical Society, Inc., 1985.

Vermont Agency of Transportation Covered Bridge Study. Prepared for the State of Vermont Agency of Transportation by McFarland-Johnson, Inc., Binghamton, New York, 1995.

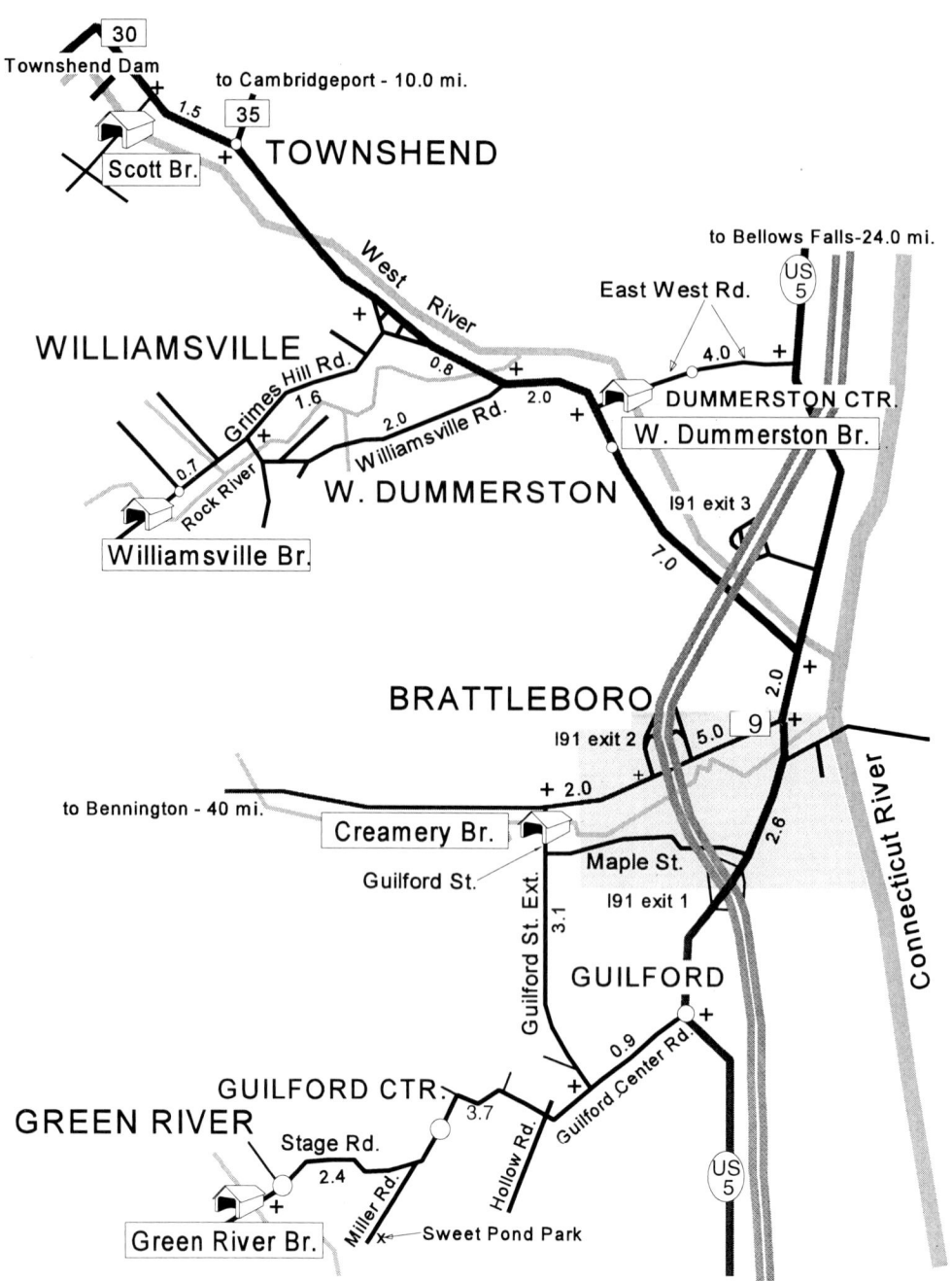

Tour 15
+ = Distance marker
Distances in miles (estimated)
Map not to scale
© 1996 Joseph C. Nelson

The DEEP SOUTH

Tour 15

There are five covered bridges to be found in the old towns of Vermont's southeast corner. All of them use Ithiel Town's plank-lattice truss, but each bridge is unique in character. The Creamery Bridge, standing next to the Molly Stark Trail in Brattleboro, may carry more traffic than any other covered bridge in the state. The two longest covered bridges in Vermont cross the West River in Townshend and in Dummerston—one of them is really three bridges connected together, and the other is two. The Williamsville Bridge in Newfane has a unique-in-Vermont gable-end treatment. The Green River Bridge in Guilford has few rivals in the beauty of its setting.

The Tour

For travelers from the Grafton area, the tour leaves Cambridgeport on Route 121 for Townshend, ten miles distant by way of Route 35. At the junction with Route 30, turn right, or west, to the Scott Bridge. After exploring the Scott, return to Townshend Village and continue on Route 30 east to Grimes Hill Road or to Williamsville Road to the Williamsville Bridge. Return to Route 30 and drive east about two miles to the West Dummerston Bridge. Continue from West Dummerston on Route 30 to the junction of U.S. Route 5 in Brattleboro. On Route 5, drive south to the junction of Route 9. Go west on Route 9 about seven miles to the Creamery Bridge.

Drive through the Creamery Bridge south on Guilford Street, then on Guilford Street Extension. Turn right on Guilford Center Road and continue west to Stage Road and, finally, the Green River Bridge. Travelers from the south can run this route backwards starting in Guilford on U.S. Route 5.

Townshend

Townshend was chartered in June 1753. Because of the French and Indian Wars, the town was not actually settled until 1761, after the area became more secure. Colonel John Hazeltine was Townshend's original grantee and he, his wife, his three daughters, and their husbands came to occupy their lands in 1769. Hazeltine had fought in the French and Indian Wars, and he was a partner with Ethan Allen in an iron foundry business in Connecticut. Perhaps this explains why Townshend was the first town east of the Green Mountains to support independence for the region that was to become Vermont.

Townshend Village stands at the junction of Route 30 and Route 35. The bridges on the West River and Saxton's River are easily accessed from here. The Scott Bridge is approximately 1.5 miles north on Route 30. Eight miles south on Route 30 find Grimes Hill Road or Williamsville Road and the Williamsville Bridge. West Dummerston Bridge lies just two miles further south.

Scott Bridge

The Scott Bridge is actually three bridges put in place by Harrison Chamberlin in 1870, after the flood of 1869. The "over-the-river" section is a 166-foot plank-lattice span. A 111-foot section, incorporating two kingpost spans supported by a pier, crosses a gully on the west bank, making the whole affair 277 feet long. The kingpost trusses use iron rods rather than the old-style timber posts.

The Scott is one of the two longest wooden bridges within the state, rivaled only by the West Dummerston Bridge. The Cornish-Windsor span is longer, but it is in two states, crossing to New Hampshire.

Probably because the plank-lattice truss was built without the customary upper secondary chords and because

Scott Bridge

TOWN: Townshend
DATE: 1870
TRUSS LENGTH: 277'3"
BUILDER: Harrison Chamberlin
TRUSS TYPE: Plank-lattice and kingpost

The Scott Bridge is reputed to stand over one of the best swimming spots on West River. Indeed, a rope is often seen hanging from the span under missing sheathing planks. The bridge was named for the former Henry Scott homestead at the west end.

of the great length of the span, the builder used four sets of buttresses to supplement the interior lateral braces and prevent the structure from twisting. The kingpost section wasn't covered until 1873, when the plank-lattice section required roof work. At that time, the town leaders had the entire bridge roofed.

The bridge was closed to traffic and deeded over to the Vermont Historic Sites Commission in 1955. A sign posted at the east portal reads:

> Longest wooden span in Vermont. This 277-foot bridge built in 1870 by Harrison Chamberlin, consists of two Kingpost trusses and a 166-foot Town lattice truss. The latter was the longest wooden span in Vermont: in 1981 a concrete pier was constructed to provide support. An earlier attempt to strengthen the bridge with the addition of a laminated bow arch was not successful.

The arch, laminated from twelve layers of three-inch plank, was added to the lattice section. The ends of the arch were bedded below the lower chords at the east abutment and at the pier supporting the kingpost section. Iron rods were suspended from the arch and bolted to the bottom bridge chords.

The arch must have proved a trial to the people responsible for the Scott's maintenance. Great care must be taken that adjustments to the iron rods supporting the weight of the bridge be done in correct order and with the correct tension. Deviation from proper "tuning" or overweight vehicles using the bridge can and did result in deformation of the arch. The reversed, or inverted, arch is still in evidence today.

While much recent work has been done to the abutments and piers, some of the original masonry can still be seen. The east bank abutment remains an example of tight and true mortarless stone construction.

Newfane

Newfane was originally chartered as "Fane" by Benning Wentworth in 1753, then again in 1761. Apparently, the area was not properly settled under these charters, so New York issued a charter to a different group in 1772, this time adding "New" to the existing "Fane."

The bridge at Williamsville is the last of Newfane's seven covered bridges. It is reached by either Grimes Hill Road or by Williamsville Road, both plainly marked where they leave Route 30 to the west. The Williamsville Road is the better of the two, as Grimes Hill Road is steep and narrow and only partially paved. The roads converge at a Rock River crossing. The covered bridge lies less than a mile further west.

Williamsville Bridge

According to DeLorme's *The Vermont Atlas and Gazetteer*, the Williamsville Bridge crosses the Rock River on Williamsville Road. Bridge historians offer a variety of different names of the stream crossed, including Stoney Brook, Marlboro Brook, and South Branch.

According to the Vermont Division of Historic Sites, the name of the Williamsville Bridge's builder and a pre-

---

Williamsville Bridge

TOWN: Newfane
DATE: c. 1870
TRUSS LENGTH: 115'6"
BUILDER: Unknown
TRUSS TYPE: Plank-lattice

The Williamsville Bridge is the last of the seven covered bridges that once graced Newfane's countryside. Historian Victor Morse wrote that the portals were painted white as a night-time aid to automobile headlights.

cise build date are unknown, but it is believed to have been built in the early 1870s, the date posted at the bridge gable-end.

Victor Morse, in *Windham County's Famous Bridges,* mentions the possibility that the bridge was either built or repaired by Caleb Lamson, builder of the Dummerston Bridge, and that it is known that the bridge survived the flood of 1869. Neal G. Templeton, on the other hand, lists the bridge as built in 1860 by Eugene P. Wheeler.

As can be expected with wooden structure of this vintage, the Williamsville Bridge has been repaired often over the years. Maybe that is why the roof gable-end overhang on the west end is plainly constructed, while the overhang on the east end is decorated with seven little corbels. The Agency of Transportation Covered Bridge Report recalls that the wooden floor beams were at one time replaced with steel beams, but these were removed in 1980 and the wooden deck restored.

The 116-foot plank-lattice span stands in a village settled in the years when homes were built close to the edge of the road. The area around the bridge is brushy and abutted by private land, limiting vantage points from which to view the bridge.

> ### Rudyard Kipling
>
> Rudyard Kipling lived in Windham County in the 1890s and may well have been a regular user of the Williamsville Bridge. He and his wife built a home they named Naulahka not far from Dummerston Village. There Kipling wrote *Captains Courageous* and the stories that became the *Jungle Book*. When he left for England in 1896, the bridge was but twenty-four years old. If the reader wants a snapshot of what Vermonters were like in those years, Kipling's "Something of Myself," a short story sometimes included in anthologies of his work, provides a revealing one.

West Dummerston Bridge

The West Dummerston Bridge, one of the two longest wooden spans in the state, stands next to busy Route 30 and a commuter parking lot. There is a little beach under the west abutment, where people come in the heat of summer to enjoy the water and the cool breezes that pass under the bridge.

Four bridges have been built over the years on this stretch of river. The Dummerston Historical Society recalls that the first bridge was lost to a spring freshet in 1826. The second bridge, built on the same abutments, was lost in 1839. The town then rebuilt downstream and lost that span in the famous flood of 1869. Caleb Lamson built the fourth and present bridge eighty rods further downstream in 1871–72. The "new" bridge was constructed with the plank-lattice truss, and it is supported in midstream on a pier.

> ### West Dummerston Bridge
>
> TOWN: West Dummerston
> DATE: 1872
> TRUSS LENGTH: 267'4"
> BUILDER: Caleb Lamson
> TRUSS TYPE: Plank-lattice

The construction of the new bridge would have started soon after the flood, but the local citizens were not able to agree on the site. After this unpropitious start, the bridge collapsed during construction. A bystander was killed by falling timbers, and Lamson was injured when he attempted to jump to safety. Since its completion, however, the West Dummerston span has served the town well for more than 120 years.

The bridge was in constant use until the middle 1990s. In 1995, it stood forlorn, bypassed by a temporary bridge. The old span was crisscrossed inside and out with a cable system designed to keep it stable while awaiting repairs in the late 1990s.

Various sources state that the Dummerston Bridge is the longer of the two longest wooden spans in the state at a length of 280 feet. Its rival, the Scott Bridge, is said to be 277 feet long. Measurements taken during the preparation of this book verified the truss length of the Scott Bridge to be 277 feet, but that of the Dummerston Bridge was found to be 267. The ridge lines measured 279 and 273 feet respectively. On the basis of these measurements, therefore, the Scott Bridge is the longest covered bridge under one continuous roof within the state, and the Dummerston is the second longest. The Dummerston span is still the longest two-span bridge, however.

The tour continues south on Route 30 to Brattleboro. About two miles south of the Dummerston bridge, the highway crosses Stickney Brook on a steel and concrete bridge. The Taft Bridge, built circa 1870, served here until it was replaced with the concrete span in 1950. The Vermont Highway Department presented the old bridge to Old Sturbridge Village in Massachusetts. Workers from Sturbridge came and dismantled the bridge and trucked it to its new home, where it has become part of a living museum.

The West Dummerston covered bridge is one of the two longest in the state, and it is the only one with diamond-shaped side ports.

BRATTLEBORO

When the border between the colonies of Massachusetts and Connecticut was settled, Massachusetts granted four tracts of land in what is now Vermont to Connecticut. The Colony of Connecticut ordered the land sold at auction in 1716. The tract that included the present towns of Brattleboro, Dummerston, and Putney went to William Dummer, William Brattle, and John White. When Massachusetts and New Hampshire settled their border in 1741, the tract became part of New Hampshire. Governor Benning Wentworth granted charters to the same proprietors in 1753, and the three towns eventually became part of the new state of Vermont. Therefore, a long-lived early settler in any of the three towns could have lived in three different states or colonies without having moved an inch.

All that remains of William Brattle's grant is his name. When the Revolutionary War broke out, he fled to Nova Scotia, and all his holdings were confiscated.

Creamery Bridge

The Creamery Bridge is located in Brattleboro on old Vermont Route 9, seven miles west of the junction with U.S. Route 5, south of the highway. Handsome in red paint, white trim, and multi-colored slate roof, the Creamery Bridge stands in an urban setting. Heavy traffic passes to the north on Route 9, and a constant flow lines up on Guilford Street to cross Whetstone Brook on the eighty-foot single-lane span.

The plank-lattice bridge was built in 1879 by A. W. Wright to replace another bridge lost to a freshet. The slate roof and the covered foot-bridge were added in 1917. The bridge is named for the Brattleboro Creamery that once stood nearby. Living Memorial Park overlooks the bridge and contributes to the ambience of the site.

The Agency of Transportation covered bridge inspection team found the bridge to be in poor condition. Because of this and because of the heavy traffic flow over the bridge, the agency has recommended that the bridge be closed and a bypass be constructed adjacent to it.

Creamery Bridge

TOWN: Brattleboro
DATE: 1879
TRUSS LENGTH: 80'1"
BUILDER: A. W. Wright
TRUSS TYPE: Plank-lattice

The Creamery Bridge carries an extremely heavy flow of traffic, and, despite its handsome appearance, shows signs of wear and tear. Note the slate roof.

GUILFORD

Guilford was chartered in 1754 by Massachusetts. On May 19, 1772, a majority of Tories and Yorkers voted that Guilford was part of Cumberland County of the Province of New York. Renouncing their charter, the townspeople governed themselves as a little republic until 1776. The town's Yorker contingent remained strong until the end of the Revolutionary War.

Green River Bridge

The 105-foot plank-lattice Green River Bridge spans the Green River surrounded by a village of the same name. The setting includes a white church on a knoll and a picturesque dam upstream.

One side of the bridge interior is lined with private mailboxes. Among them, distinctively red, white, and blue, is the only U.S. Postal Service box in a bridge in the state. Look up into the roof rafters to see where the road crew stores the town's gin pole.

The red-painted portals are rounded and trimmed with a molding. The gables are finished with horizontal butted planks. A sign warns about a two-dollar fine if a user passes through at a pace faster than a walk. The bridge was built in 1870 by Marcus Worden, replacing a span that was lost in the 1869 flood.

The Green River Bridge lies at the end of Guilford Center Road and Stage Road. Guilford Center Road is easily reached by driving through the Creamery Bridge on Guilford Street, then on Guilford Street Extension, which ends on Guilford Center Road. Turn right to go to the bridge.

Guilford Center Road twists through a maze of other little roads, marked and unmarked. The traveler needs to have faith and count road entries and exits carefully on the map and rejoice when a corroborating road sign is seen. Stage Road is narrow, very hilly, and unpaved, with barely two-car passage. In summer, it is a green tunnel; in the fall, the trees are a riot of color. Casual passage in winter or in mud season is not recommended.

Green River Bridge

TOWN: Green River
DATE: 1870
TRUSS LENGTH: 104'6"
BUILDER: Marcus Worden
TRUSS TYPE: Plank-lattice

Sheltering mail boxes and standing next to a park, the Green River Bridge is at the center of village life in Green River.

Ethan Allen and the Guilford Yorkers

In 1782 pro-New York Guilford residents voted to stand against the "pretend-state" of Vermont. They believed that with the war ended, Congress would move against Ethan Allen. Until then, they voted to receive direction from New York's Governor Clinton. The Yorkers and Vermonters in town carried on with hostile parallel governments.

In August 1782 Ira Allen sent a sheriff to arrest the commander of the Yorker Militia. The Yorkers resisted, and the sheriff came away empty-handed. Governor Thomas Chittenden then ordered Ethan Allen to enforce Vermont's authority. Allen rode at the head of a hundred of his men to Guilford. There he gave a bombastic speech in which he promised no quarter to any who opposed him, and unless they accepted his authority, he would "lay Guilford as desolate as Sodom and Gomorrah!" The Yorkers ran away. Twenty of them were captured and taken to a grand jury in Westminster, where all but four of them agreed to join Vermont. The four holdouts lost their property and were sent to New York.

Sources

Allen, Richard Sanders. *Covered Bridges of the Northeast.* Brattleboro, VT: Stephen Greene Press, 1957.

Bellesiles, Michael A. *Revolutionary Outlaws: Ethan Allen and the Struggle for Independence on the Early American Frontier.* Charlottesville, VA: University Press of Virginia, 1993.

Brodie, Jocelyn, et al. *Townshend and the Founding of Vermont.* Brattleboro, VT: West Townshend Historical Society, 1991.

Congdon, Herbert Wheaton. *The Covered Bridge.* Middlebury, VT: Vermont Books, 1973.

Dummerston: An "Equivalent Lands" Town 1753-1986. Dummerston, VT: Dummerston Historical Society, 1990.

Hemenway, Abby Maria. *The Vermont Historical Gazetteer, Vol. V.* Burlington, VT: A. M. Hemenway, 1891.

Morse, Victor. *Windham County's Famous Bridges.* Brattleboro, VT: Stephen Greene Press, 1960.

Vermont Agency of Transportation Covered Bridge Study. Prepared for the State of Vermont Agency of Transportation by McFarland-Johnson, Inc., Binghamton, New York, 1995.

A SUMMARY of VERMONT'S COVERED BRIDGES

Appendix A

One-hundred-and-two of Vermont's covered bridges are herein charted and summarized by tour. The bridges are counted by truss type and status. Superlatives—longest, shortest, highest, lowest, oldest, and so on—are noted, along with other unique features.

Explanation of Terms

Bridge name. The name used in this text. Other names for the bridge are listed under Comments.

Builder and Build date. The name of the bridge builder and the year the bridge was built are listed when known. These are sometimes gleaned from town records, old newspapers, and folklore. There are often disagreements about build dates among historians and engineers—these are mentioned in the text section.

Location. The name of the host municipal unit or village.

Truss type. The bridge truss used is classified as one of several general truss types.

Stream. The name of the stream the bridge crosses, according to the Vermont Tour Guide map.

Orientation. The cardinal magnetic compass direction along which the bridge ridge pole lies. This is meant to be an aid to the photographer in predicting where the sun will be during the day and season relative to the chosen view of the subject bridge.

Status. The bridge is listed as carrying all traffic (subject to the weight limit signs), carrying foot and cycle traffic only, or retired.

Deck reinforced. When the bridge structure is considered no longer safe, the bridge deck is made self-supporting by means of steel beams under the old deck or by replacing the old floor system with a concrete slab. When this is done, the historic bridge truss no longer supports traffic and the bridge is considered by some as no longer being a true covered bridge.

Measurements. To compile this summary, the author visited, studied, and measured each bridge. These measurements were taken because this writer found many of the published bridge lengths to be wildly at odds with reality.

> **Truss length.** Ideally, the length of the bridge from end post to end post. Practically, it is the length of the bridge from portal face to portal face.
>
> **Portal.** The inside width of the portal entry, ignoring the deck width. The measurements were taken to discover the average portal width and the bridges with the narrowest and the widest portals.
>
> **Over-all-length.** The length of the bridge roof at the ridge pole. Practically, the measurement is the sum of the truss length, the gable end overhang, and the roof gable end overhang.
>
> **Deck.** The length of the bridge roadway measured from abutment back wall to abutment back wall.

NOTE: Notice that some of the notes are preceded by the notation VAOT#. These are the seventy-five bridges included in the Vermont Agency of Transportation covered bridge study.

Build Date	Truss Type	Truss Length	Portal (W/H)	Overall Length	Builder

TOUR ONE—BENNINGTON COUNTY BRIDGES

BRIDGE-AT-THE-GREEN

| 1852 | Lattice | 80'4" | 14'1"/12'5" | 83' | unknown |

VAOT #24. Serves as entry-way to green. Stands next to chapel, mature trees, swimming hole. Renovated 1978–80. Remains structurally original.

CHISELVILLE

| 1870 | Lattice | 116'6" | 11"/10'11" | 116'6" | Daniel Oatman |

VAOT #6. Supported with guy wires. Roadway reinforced with steel beams under the roadway, midstream pier added 1973. At 40 feet, second highest above stream bed in Vermont. *Of the Batten Kill.

HENRY

| 1989 | Lattice | 120'6" | 14'9"/11'1" | 123'2" | VAOT |

VAOT #32. Replica of circa 1840 span constructed by VAOT in 1989. Original span famous for doubled truss, not replicated.

PAPER MILL

| 1889* | Lattice | 125'2" | 14'9"/11'2" | 127'2"' | Charles F. Sears |

VAOT #31. Other name(s): Bennington Falls Bridge. Bypassed by temporary one-lane bridge. The bridge stands over a mill pond and dam, next to the old brick mill buildings. The truss plank web is repaired using "sisters." Closed to all traffic: inspection by VAOT in 1993 found bridge to be on verge of collapse. Restoration is planned for 1998–2000. *c. 1840 according to J. Spargo of the Bennington Museum.

SILK ROAD

| c. 1840* | Lattice | 88'3" | 15'/11'11" | 55' | Benjamin Sears* |

VAOT #30. Other name(s): Locust Grove Bridge, Robinson Bridge. N.G. Templeton gives a date of 1889. The plank web was repaired using "sisters," roofed with wood shingles, additional wood floor beams placed in structure, dates of work unknown. Restored in 1991 by Gilbert Newbury of the VAOT. * according to J. Spargo of the Bennington Museum.

TOUR TWO—THE OTTER CREEK BASIN

BROWN

| 1880 | Lattice | 112'6" | 15'/11'7" | 112'6" | Nicholas Powers |

VAOT #34. Nicholas Powers' last project. Iron tie rods between bottom chords for reinforcement. Siding stops short of eaves. Roof shingled with slate. Very strong. One of best preserved Town lattice bridges in Vermont (most original members) per Gilbert Newbury. Crosses good trout stream. Abutment stands on boulder.

COOLEY

| 1849 | Lattice | 50'3" | 15'/12' | 67'7" | Nicholas Powers |

VAOT #31. Features exaggerated extended gable ends.

Location	Stream	Orientation	Status Capacity	Deck Girders?
Arlington	Batten Kill	N-S	in use / 8T	no
Sunderland	Roaring Branch*	NE-SW	in use / 25T	yes
Bennington	Walloomsac R.	NE-SW	in use / 5.5T	no
Bennington	Walloomsac R.	NE-SW	in use / 20T	no
Bennington	Walloomsac R.	N-S	in use / 9T	no
Shrewsbury	Cold River	NW-SE	in use / 9T	no
Pittsford	Furnace Brook	NE-SW	in use / 8T	no

Build Date	Truss Type	Truss Length	Portal (W/H)	Overall Length	Builder

DEPOT
| 1840 | Lattice | 120'6" | 18'/12'1" | 133'6" | Abraham Owen |

VAOT #33. Supported by bar anchored in ground to top of truss on SE side. Reinforced with four steel beams and supplemental piers under the roadway. Roof shingled with slate.

GORHAM
| 1841 | Lattice | 114'2" | 17'9"/12'4" | 129'4" | A. Owen, N. Powers |

VAOT #4. Other Name(s): Goodnough Bridge. Distribution beams added under the deck.

HALPIN
| 1824 | Lattice | 66'4" | 12'/11'5" | 66'4" | unknown |

VAOT #23. Serves farm on dead-end road. At 41 feet, highest bridge above stream in state (VT Historic Div. has it 46 feet). Reconstructed in 1994. Once noted for very tall cut marble abutments, now replaced with cast concrete. Marble-stone foundations of mill or store can be seen at east end. *Muddy Branch of the New Haven River.

HAMMOND
| 1842 | Lattice | 138'8" | 17'10/ - | 152' | Asa Nourse |

Historic Site Marker: listed as Vermont Historic Site.

KINGSLEY
| 1836* | Lattice | 120'6" | 14'1"/12'11" | 120'6" | T.K. Norton |

VAOT #28. Other name(s): Mill River Bridge. Serves East Street. Stands downstream from falls and mill at East Clarendon. Steel cables on all four corners. Closed for repair in 1950 and again in 1985, reopened in May 1987. *per town history. Other sources offer 1870.

PULP MILL
| 1820* | Burr Arch | 183'7" | 11'7"/10'1" | 199'3" | unknown |

VAOT #1. Built for Waltham Turnpike Co.(per H.W. Congdon) Single span, two lane bridge with three flanking arches. Renovated in 1979, arches modified, two sets of timber cribs on masonry piers added under roadway. When the timber arches were removed and laminated arches with iron rod hangers substituted, technically, this is no longer a Burr Arch bridge. *Date per H.W. Congdon. Jan Lewandoski suggests 1850s, per Selectman's records.

RUTLAND RAILROAD
| 1897 | Howe | 109' | 13'10"/ - | 132' | Rutland R.R. |

Other name(s): Shoreham Railroad Bridge. The railroad track has been lifted. The bridge on the Register of Historic Sites.

SALISBURY STATION
| 1865 | Lattice | 153'10" | 13'8"/13'3" | 155'6" | unknown |

VAOT #8. Other name(s): Station Bridge, Cedar Swamp Bridge. Widest spaced plank lattice in state (4 ft. 10 in. vs 3-ft. average). Center pier added in 1969. Jan Lewandoski made repairs in winter of 1992.

Location	Stream	Orientation	Status Capacity	Deck Girders?
Pittsford	Otter Creek	NE-SW	in use / 25T	yes
Pittsford-Proctor	Otter Creek	NE-SW	in use / 3T	no
Middlebury	Muddy Branch*	E-W	in use / 20T	no
Pittsford	Otter Creek	E-W	retired	no
East Clarendon	Mill River	N-S	in use / 3T	no
Middlebury & Weybridge	Otter Creek	E-W	in use / 4T	yes
Shoreham	Lemon Fair River	N-S	retired	no
Cornwall & Salisbury	Otter Creek	E-W	in use / 7T	yes

A SUMMARY OF VERMONT'S COVERED BRIDGES

Build Date	Truss Type	Truss Length	Portal (W/H)	Overall Length	Builder

SANDERSON

c. 1838*	Lattice	132'	18'/12'7"	132'8"	unknown

VAOT #12. One of only two bridges in Vermont with enclosed cornices and cornice returns. Closed and bypassed with a temporary bridge 1984. *Per N. G. Templeton. VAOT report lists c. 1840 build date.

TWIN

1850	Lattice	64'6"	15'1"/ -	66'10"	Nicholas Powers

*Converted by the Town of Rutland to a barn after being washed away by flood when the East Creek dam broke in 1957. Other twin was destroyed in the disaster.

TOUR THREE—THE WOODEN BRIDGES OF CHARLOTTE

LAKE SHORE

1898*	Tied arch**	40'1.5"	11'11"/11'5"	43'1.5"	Leonard Sherman

VAOT #27. Other name(s): Holmes Creek Bridge. One of three surviving tied-arch truss bridges. The others are the Bowers and Bests bridges in West Windsor. Extensive repairs were made in 1993 by Graton associates. *Date unknown; AOT assumes 1898. **Tied arch with kingpost. The kingpost truss serves no function.

MUSEUM

1845	Burr arch	156'	26'6"*/ -	169'	Farewell Wetherby

Other name(s): Big Bridge while in Cambridge. Moved from Cambridge in 1951, the bridge is now owned by the Shelburne Museum and currently in use as a private entrance to the museum grounds. One of only two two-lane covered bridges in the state. *The portal of the north lane is 12'11" wide; the south is 12'9" wide. The span is constructed with three parallel segmented timber arches with bases embedded in the abutments below the main chords.

QUINLAN

1849	Burr arch	85'6"	13'6"/11'11"	92'6"	unknown

VAOT #29. Other name(s): Lower Bridge, Sherman Bridge. Strengthened with two auxiliary steel beams under existing floor beams. Distribution beam added. Unusual lateral roof bracing.

SEGUIN

1849	Burr arch	70'3"	13'4"/12'	78'3"	unknown

VAOT #28. Other name(s): Upper Bridge. Unusual lateral roof bracing. Extensive repairs in 1949 and 1994. There are scenic waterfalls upstream.

SPADE FARM

1824*	Lattice	85'6"	17'/ -	93'6"	Justin Miller

The Spade Farm Bridge is privately owned. The span once crossed the Otter Creek—it was moved to its present location in 1953. It is in dangerous condition. *R. S. Allen dates bridge c. 1850.

Location	Stream	Orientation	Status Capacity	Deck Girders?
Brandon	Otter Creek	E-W	retired	no
Rutland	East Creek*	E-W	retired	no
Charlotte	Holmes Creek	NE-SW	in use / 3T	no
Shelburne	Burr Pond	N-S	private entry	no
Charlotte	Lewis Creek	NW-SE	in use / 12T	yes
Charlotte	Lewis Creek	SW-NE	in use / 5T	no
No. Ferrisburgh	unnamed pond	SE-NW	private	no

Build Date	Truss Type	Truss Length	Portal (W/H)	Overall Length	Builder

TOUR FOUR—THE LAMOILLE RIVER AND THE NORTH BRANCH

CAMBRIDGE JUNCTION

1887	Burr arch	140'	15'8"/11'6"	160"	G.W. Holmes

VAOT #29. Other name(s): Poland Bridge. Distribution beams added under deck. Serves Junction Road (TH #23). *Closed by VAOT. Restoration funds being sought. Has longest clear-span of any Burr truss bridge in Vermont.

CHURCH STREET

c. 1877*	Queenpost	60'	12'5"/10'11"	61'	unknown

VAOT #14. Other name(s): Village Bridge. Reconstructed in 1968 with independent roadway reinforced with four steel beams. *Build date undetermined; VAOT dates bridge 1895, per Hagerman, possibly 1877. **North Branch of the Lamoille River.

EAST FAIRFIELD

1865	Queenpost	68'	13"/10'5"	68'	unknown

VAOT #50. *Closed to vehicular traffic, in use as foot bridge. The bridge stands over a mill pond.

GATES FARM

1897	Burr Arch	82'	16'3"/11'	92'3"	G. W. Holmes

Other name(s): Little Bridge. Serves Gates Farm. The original bridge was built by George Washington Holmes in 1897—it was moved to present location in 1950. Arches butt on chords. Due to severe rot, the bridge underwent extensive renovation in 1994. The plans were designed by Gilbert Newbury.

GRIST MILL

unknown	Burr Arch	87'7"	14'9"/11'	97'7"	unknown

VAOT #30. Other name(s): Scott Bridge, Bryant Bridge, Canyon Bridge. Arches end at chords. Distribution beams added under deck.

JAYNES

c. 1877	Queenpost	57'	11'10"/12'4"	62'	unknown

VAOT #15. Other name(s): Upper Bridge, Codding Hollow Bridge, Kissing Bridge. Reconstructed in 1960 making independent roadway reinforced with four steel beams. *North Branch of the Lamoille River.

LUMBER MILL

c. 1895*	Queenpost	70'7"	11'10/12'4"	72'6"	Lewis Robinson

VAOT #12. Other name(s): Mill Bridge, Junction Bridge, Lower Bridge. Reinforced in 1971–72 with four steel beams under roadway. Timbers in chords and queenpost braces replaced in 1995 by Paul Ide and bridge restorer Jan Lewandoski. *Probable build date believed to be mid 1890s. ** North Branch of the Lamoille River.

MAPLE STREET

1865	Lattice	56'6"	17'/10'6"	56'6"	Kingsbury, Stone

VAOT #25. Other name(s): Lower. The plank web has been repaired with "sisters." Extensive rehabilitation done in 1990-91. One of widest covered bridge roadways in Vermont.

Location	Stream	Orientation	Status Capacity	Deck Girders?
Cambridge	Lamoille River	N-S	status*/ 3T	no
Waterville	North Branch**	E-W	in use / 20T	yes
Fairfield	Black Creek	NE-SW	foot bridge*	no
Cambridge	Seymour River	NW-SE	serves farm	no
Cambridge	Brewster Brook	E-W	in use / 5T	no
Waterville	North Branch*	S-N	in use / 18T	yes
Belvidere	North Branch**	N-S	in use / 30T	yes
Fairfax	Mill Brook	NE-SW	in use / 3T	no

Build Date	Truss Type	Truss Length	Portal (W/H)	Overall Length	Builder

MONTGOMERY
| 1887* | Queenpost | 63' | 12'/10'8" | 70'6" | unknown |

VAOT #16. Other name(s): Lower Bridge, Potter Bridge. Named for Dallas Montgomery farm. Reconstructed in 1971 with independent roadway reinforced with four steel beams. *Possibly 1887, per R. L. Hagerman. **North Branch of the Lamoille River.

MORGAN
| 1887 | Queenpost | 63'8" | 12'2"/11'6" | 73' | Robinson, Tracy, & Leonard |

VAOT #13. Other name(s): Upper Bridge. Remarkable innovations in queenpost truss.
** North Branch of the Lamoille River.

WESTFORD
| 1838 | Burr arch | 96'6" | 13'6"/ - | 96'6" | unknown |

Other name(s): Browns River Bridge. Repaired and waiting for return to abutments. Planned for foot and bicycle traffic only.

TOUR FIVE—THE TOWN OF MONTGOMERY

COMSTOCK
| 1883 | Lattice | 68'10" | 16'4"/11'4" | 71'10" | S. & S. Jewett |

VAOT #41. Only Jewett bridge with side port. Distribution beams added under deck. Foundations of Comstock carriage factory upstream.

CREAMERY
| 1883 | Lattice | 58'8" | 16'3"/11'4" | 61'8" | S. & S. Jewett |

VAOT # 32. Other name(s): West Hill Bridge, Crystal Springs Bridge. The bridge stands over a natural waterfall and pool. The foundations of the old creamery are at the east end of the bridge. *Closed after VAOT inspection in 1994.

FULLER
| 1890 | Lattice | 49'8" | 16'6"/11'4" | 52'7" | S. & S. Jewett |

VAOT #5. Other name(s): Black Falls Bridge.

HECTORVILLE
| 1883 | Lattice* | 52'9" | 16'/ - | 55'9" | S. & S. Jewett |

VAOT #31. Closed and bypassed with concrete bridge. Upstream, the river winds through rocks to pool, cascades downstream. *Bridge strengthened with "jury-rigged" kingpost truss. **South Branch of the Trout River.

HOPKINS
| 1875 | Lattice | 90'9" | 16'/11'4" | 94'9" | S. & S. Jewett |

VAOT #52. *Closed to traffic in 1995 after inspection by VAOT. Reconstruction slated for 1998–2000. Currently bypassed with portable bridge. Previous repairs included two steel beams cantilevered from west abutment to support ends of bottom chords. Longest of Jewett bridges.

Location	Stream	Orientation	Status Capacity	Deck Girders?
Waterville	North Branch**	S-N	in use / 20T	yes
Belvidere	North Branch**	N-S	in use / 5T	no
Westford	Brown's River	SW-NE	retired	no
Montgomery	Trout River	E-W	in use / 3T	no
Montgomery	West Hill Brook	SW-NE	status*	no
Montgomery	Black Falls Creek	N-S	in use / 3T	no
Montgomery	South Branch**	E-W	retired	no
Enosburg	Trout River	E-W	status*/ -	no

Build Date	Truss Type	Truss Length	Portal (W/H)	Overall Length	Builder

HUTCHINS

1883	Lattice	77'	16'/11'2"	80'	S. & S. Jewett

VAOT #34. Waterfalls below bridge. Foundations of butter tub mill near east portal. Serves dead end road. *South Branch of the Trout River.

LONGLEY

1863	Lattice	84'7"	16'/11'4"	87'7"	S. & S. Jewett

VAOT #33. Other name(s): Samuel Head Bridge, Harnois Bridge. Reinforced with three steel beams cantilevered from west abutment to end of stringers, abutment work: 1979. New deck, additional floor beams, chord repairs, siding and roof repairs in 1992–93 by Jan Lewandoski.

TOUR SIX—ROUTE 100 IN NORTHERN VERMONT

BLACK RIVER

1881	Paddleford	85'10"	14'10"/ -	94'4"	John D. Colton

VAOT #20. Other name(s): Orne Bridge, Lower Bridge, Coventry Bridge. One of three Paddleford truss bridges surviving in Vermont.

FISHER

1908	Town-Pratt	103'	15'/ -	109'	Pratt Constr. Co.

Railroad bridge. Features full length roof vent for steam trains. Steel beams and a mid-stream pier added under roadbed. *Railroad currently inactive, bridge is useable.

LORDS CREEK

1881	Paddleford	47'9"	15'/11'8"	57'3"	John D. Colton

Named for stream it was moved from. One of three Paddleford truss bridges surviving in Vermont. Serves farm field.

POWER HOUSE

1870	Queenpost	60'10"	16'2"/11'10"	72'10"	unknown

VAOT #4. Other name(s): School Street Bridge.

SCRIBNER

1919*	Queenpost	44'6"	13'7"/12'	47'9"	unknown

VAOT #30. Other name(s): Mudget Bridge. Unique half-high queenpost truss. Roadway reinforced with steel beams in 1960. * Date unknown; VAOT assumes 1919.

SCHOOL HOUSE

1910	Lattice	92'4.5"	12'5"/9'8"	99"	unknown

VAOT #8. Other name(s): River Road Bridge, Upper Bridge. The lattice truss lacks secondary chords: one of the few Vermont bridges still featuring buttresses. Uses single chord pieces on each side of the lattice members instead of usual two. Also unique in that lattice truss crossings are pinned with single treenails.

Location	Stream	Orientation	Status Capacity	Deck Girders?
Montgomery	South Branch*	E-W	in use / 3T	no
Montgomery	Trout River	E-W	in use / 9T	no
Irasburg	Black River	NW-SE	in use / -	no
Wolcott	Lamoille River	N-S	functional*	yes
Irasburg	Black River	E-W	private	no
Johnson	Gihon River	E-W	in use / 3T	no
Johnson	Gihon River	SW-NE	in use / 20T	yes
Troy	Missisquoi R.	E-W	in use / 8T	no

A SUMMARY OF VERMONT'S COVERED BRIDGES

Build Date	Truss Type	Truss Length	Portal (W/H)	Overall Length	Builder

TOUR SEVEN—CROSSING THE CONNECTICUT

COLUMBIA

| 1912 | Howe | 146" | 15'/12'11" | 148' | Charles Babbitt |

Owned by the State of New Hampshire. The single-lane Columbia Bridge crosses the Connecticut in one free span. *Posted limit.

CORNISH-WINDSOR

| 1866 | Timber Lattice | 460' | - /9'3" | 476' est. | Tasker, Fletcher |

Owned by the State of New Hampshire. Two spans with undivided two lanes of traffic. Extensive renovations in 1988–89—timber work done by Jan Lewandoski. *Posted limit.

MOUNT ORNE

| 1911 | Howe | 266'11" | 15'6"/13' | 269'5" | Charles Babbitt |

Other name(s): Lunenburg-Lancaster Bridge. Owned by the State of New Hampshire. The single-lane Mount Orne Bridge crosses the Connecticut in two spans. *Posted limit.

TOUR EIGHT—LYNDON

CHAMBERLIN

| unknown | Queenpost | 65'11" | 16'5"/12'5" | 85' | W.W. Heath |

VAOT #33. Other name(s): Chamberlin Mill Bridge, Sawmill Bridge, Whitcomb Bridge. Roof added to the existing open bridge in 1881. Open sides, peculiar to Lyndon area. *South Wheelock Branch of the Passumpsic River.

GREENBANKS HOLLOW

| 1886 | Queenpost | 74'5" | 14'11"/12'4" | 75' | unknown |

VAOT #40. Open sides, similar to bridges in Lyndon area. Stands below stone mill foundations and above broken dam. Reinforced with two above-deck steel beams and a pier in 1970. *Posted limit.

MILLER'S RUN

| 1995 | Queenpost | 53'8" | 15'1"/ - | 71'8" | E.H. Stone |

Other name(s): Bradley Bridge. Open sides, peculiar to Lyndon area. Originally built in 1878, the bridge was reconstructed by the VAOT in 1995. About half of the original truss was retained.

RANDALL

| 1865 | Queenpost | 67'2" | 12'6"/ - | 79' | unknown |

Other name(s): Barrington Road Bridge. Retired, in use as a footbridge, part of snowmobile trail. Open sides, peculiar to Lyndon area. Private bridge. *East Branch of the Passumpsic River.

SANBORN

| 1869 | Paddleford | 120' | 17'10"/ - | 132' | unknown |

Other name(s): Center Bridge. Open sides, peculiar to Lyndon area. One of three Paddleford truss bridges surviving in Vermont. Features covered walkway. Owned by motel. The bridge was replaced by a modern span. Moved to its present location in 1960 by Milton Graton.

Location	Stream	Orientation	Status Capacity	Deck Girders?
Lemington	Connecticut River	NW-SE	in use / 6T*	no
Windsor	Connecticut River	N-S	in use / 10T*	no
Lunenburg	Connecticut River	SE-NW	in use / 6T*	no
Lyndon Corner	S. Wheelock Branch*	SW-NE	in use / 5T	no
Danville	Joe's Brook	N-S	in use / 6T*	yes
Lyndon Center	Millers Run	N-S	in use / 20T	yes
Lyndonville	East Branch*	N-S	foot bridge	no
Lyndon Center	Passumpsic R.	SW-NE	retired	no

A SUMMARY OF VERMONT'S COVERED BRIDGES

Build Date	Truss Type	Truss Length	Portal (W/H)	Overall Length	Builder

SCHOOL HOUSE

1879　　Queenpost　　41'8"　　16'3"/ -　　55'　　Jones, Goodell

One of only two covered bridges in Vermont with completely enclosed trusses. Originally had two covered walkways, now has one. Stands in town green, next to commuter parking. *South Wheelock Branch of the Passumpsic River.

TOUR NINE—ROUTE 100 IN CENTRAL VERMONT

BIG EDDY

1833*　　Burr arch　　105'1"　　16'2"/10'4"　　112"　　unknown

VAOT # 4. Other name(s): Village Bridge, Great Eddy Bridge. Restored by Milton Graton in 1973. Features attached foot bridge. Knee braces cut from tree branches or roots. Distribution beams added under deck. Floor redecked in 1989, structural repairs made in 1992. A bedrock outcrop supports north abutment and creates eddy bridge is named for. *N. G. Templeton offers date of 1853.

COBURN

1851　　Queenpost　　50'　　13'5"/11'3"　　70'4"　　Larned Coburn

VAOT #22. Other name(s): Cemetery Bridge. Renovated in 1972–73—deck replaced with cast concrete supported by steel beams. Truss renovated winter 1996–97.

GOLD BROOK

1844　　Howe　　48'5"　　13'9"/11'4"　　56'5"　　John W. Smith

VAOT #49. Other name(s): Stowe Hollow Bridge, Emily's Bridge. Reputed to be haunted by Emily's ghost. Once unique for railroad rails supporting deck, since removed. Only Howe truss highway bridge within Vermont. Residents directed selectmen to provide perpetual maintenance (Hagerman).

LINCOLN GAP

1880　　Queenpost　　54'11"　　13'11"/12'9"　　64'7"　　Walter Bagley

VAOT #6. Other name(s): Warren Bridge. Located within Historic Residential District. One of only two covered bridges in Vermont with enclosed trusses. The roadway strengthened with timber cross-braces and stringers and roof repaired in 1995 by Paul Ide for bridge restorer Jan Lewandoski. The bridge stands upstream of a mill-dam and waterfall.

MARTIN

1890　　Queenpost　　44'9"　　11'/ -　　52'1"　　H.F. Townsend

Other name(s): Orton Farm Bridge. The Martin Bridge is impassible—its deck lacks ramps, and the floor-planking is unsound.

PINE BROOK

1872　　Kingpost　　48'6.5"　　14'2"/12'　　50'6.5"　　unknown

VAOT #20. Other name(s): Wilder Bridge. Serves North Road. Restored by Milton Graton in 1976. Two steel beams aligned under deck to assist bridge structure only in event of overload. Floor re-decked in 1989. One of the two kingpost truss bridges surviving with wood kingposts.

Location	Stream	Orientation	Status Capacity	Deck Girders?
Lyndon Corner	S. Wheelock Branch*	SE-NW	retired	no
Waitsfield	Mad River	SW-NE	in use / 3T	no
E. Montpelier	Winooski River	NW-SE	in use / 25T	yes
Stowe	Gold Brook	NE-SW	in use / 3T	no
Warren	Mad River	E-W	in use / 8T	no
Marshfield	Winooski River	E-W	private	no
Waitsfield	Pine Brook	NE-SW	in use / 18T	yes

Build Date	Truss Type	Truss Length	Portal (W/H)	Overall Length	Builder

RED

| 1896 | Unique truss | 64'2" | 15'3"/11'10" | 74'2" | unknown |

VAOT #39. Other name(s): Sterling Brook Bridge, Chaffee Bridge. The truss appears to be an outsized kingpost with superimposed queenpost truss. The wooden deck was replaced with a cast concrete roadway and two steel beams in 1971. Bedrock provides natural abutments.

TOUR TEN—NORTHFIELD

MOSELEY

| 1899* | Kingpost | 36'6" | 15'9"/12'11" | 39'6" | John Moseley |

VAOT #63. Other name(s): Stoney Brook Bridge. One of two surviving kingpost bridges with timber kingposts. Reconstructed in 1971 with an independent timber deck supported by five steel beams. Once known for the large granite stones used, the abutments were faced and capped with concrete in 1990. Siding repaired and painted in 1978. *N. G. Templeton dated the bridge 1899.

NEWELL

| 1872 | Queenpost | 56'5" | 15'3"/12'8" | 59'5" | unknown |

VAOT #11. Other name(s): Lower Cox Brook Bridge. Only location in Vermont from which three covered bridges can be seen; the others are the Station and Upper Cox Brook bridges. Strengthened in the 1960s with an independent timber deck supported by four steel beams.

SLAUGHTER HOUSE

| c. 1876* | Queenpost | 59'7" | 11'9"/10'10" | 62'7" | unknown |

VAOT #64. Serves "one-house" dead end road. Only covered bridge in Northfield Falls not altered structurally. Supported by abutments of unmortared stone blocks and slabs. *VAOT covered bridge report offers build date of 1872. **Posted limit.

STATION

| 1872 | Lattice | 136'10" | 16'2"/12'8" | 147'2" | unknown |

VAOT #15. Other name(s): Northfield Falls Bridge. Newell Bridge can be seen from here. Strengthened in 1963 with an independent timber deck supported by four steel beams, center pier added. Tie rods extend between top chords for lateral bracing. Guy cables extend to river bank from upper SE & SW corners. Siding repaired and painted in 1978.

UPPER COX BROOK

| c. 1872* | Queenpost | 51'3" | 13'11"/12'6" | 54'3" | unknown |

VAOT #10. Other name(s): Upper Bridge. Strengthened in 1966 with an independent timber deck supported by four steel beams. Siding repaired in 1978. *VAOT covered bridge report offers c. 1872.

TOUR ELEVEN—THE NORTHERN TRIBUTARIES OF THE WHITE RIVER

CILLEY

| 1883 | Multiple Kingpost | 59'10" | 15'3"/10'8" | 71'4" | unknown |

VAOT #33. Other name(s): Lower Bridge. The stream crossing is skewed.

Location	Stream	Orientation	Status Capacity	Deck Girders?
Morristown	Sterling Brook	NE-SW	in use / 25T	yes
Northfield	Stoney Brook	NW-SE	in use / 4.5T	yes
Northfield	Cox Brook	E-W	in use / 19T	yes
Northfield	Dog River	E-W	in use / 16T**	no
Northfield	Dog River	E-W	in use / 32.5T	yes
Northfield	Cox Brook	NW-SE	in use / 25T	yes
Tunbridge	White River, 1st Branch	N-S	in use / 3T	no

Build Date	Truss Type	Truss Length	Portal (W/H)	Overall Length	Builder

FLINT

1845	Queenpost	87'7"	15'/11'5"	90'	unknown

VAOT #35. Located near the Justin Morgan farm historic site. Restored in 1969. Noted in VAOT inspection report as "Outstanding example of functional preservation of an historic structure." Interesting tension splice in the bottom chords.

GIFFORD

1904	Multiple Kingpost*	51'8"	12'3"/10'9"	60'	unknown

VAOT #34. Other name(s): C. K. Smith Bridge, Blue Bridge(it was once painted blue). Reinforced with two steel beams above roadway tie-rodded to smaller steel beams under deck. * Half-high.

HOWE

1879	Multiple Kingpost	74'4"	13'1"/11'8"	77'	Mudget, Tenny, and Wells

VAOT #32. Leads traffic past Howe farm. Repairs made in 1994, including deck and floor beam replacement.

HYDE

1904	Multiple Kingpost	51'8"	15'9"/12'	53'8"	unknown

VAOT #40. Other name(s): Kingsbury. Marked "So. Randolph Vt." Serves farm. Repaired in 1994.

JOHNSON

1904	Multiple Kingpost*	37'3"	14'/10'6"	39'7"	unknown

VAOT #38. Other name(s): Upper Blaisdel Bridge, Braley Bridge. Strengthened in 1977 with four steel beams under roadway. Began as an open bridge, covered in 1909 (Congdon). * Half-high.

LARKIN

1902	Multiple Kingpost	67'8"	13'/9'10"	72'	Arthur Adams

VAOT #34. Note stream crossing is skewed.

MILL

1883	Multiple Kingpost	72'	15'6"/12'	78'	unknown

VAOT #5. Other name(s): Hayward and Noble Bridge. Stands below mill dam in village mill district.

MOXLEY

1883	Queenpost	53'10"	14'7"/10'8"	63'10"	Arthur Adams

VAOT #46. Other name(s): Guy Bridge. Very long hand-hewn chords. Wooden ramp at north end. Stream crossing is skewed.

Location	Stream	Orientation	Status Capacity	Deck Girders?
Tunbridge	White River, 1st Branch	E-W	in use / 5T	no
Randolph	White River, 2nd Branch	E-W	in use / 3T	yes
Tunbridge	White River, 1st Branch	E-W	in use / 10T	no
Randolph	White River, 2nd Branch	E-W	in use / 5.5T	no
Randolph	White River, 2nd Branch	E-W	in use / 25T	yes
Tunbridge	White River, 1st Branch	NW-SE	in use / 6T	no
Tunbridge	White River, 1st Branch	NW-SE	in use / 8T	no
Chelsea	White River, 1st Branch	N-S	in use / 4T	no

A SUMMARY OF VERMONT'S COVERED BRIDGES

Build Date	Truss Type	Truss Length	Portal (W/H)	Overall Length	Builder

TOUR TWELVE—WOODSTOCK

LINCOLN

1865*	Pratt-arch	135'10"	14'7"/10'11"	141'6"	R.W. & B.H. Pinney

VAOT #25. Serves Fletcher Hill Road. Pratt truss modified by addition of arch. Only known example of Pratt-arch truss implemented in wood in U.S. Completely renovated in 1947. Distribution beams added under deck. Renovated in 1989 by Wright Construction Co. *R. S. Allen gives build date of 1877.

MIDDLE

1969	Lattice	125'	14'9"/14'	139'	Milton Graton

VAOT #73. Other names(s): Union Street Bridge. Built by Milton Graton in 1969. Fire damage repaired in 1976.

SAYERS

unknown	Haupt, with arch	129'	18'8"/11'5"	130'	unknown

VAOT #27. Other name(s): Thetford Center Bridge. On the National Register of Historic Places. Only Haupt truss-type bridge in New England. Natural waterfalls downstream. The bridge roadway was reinforced with four steel beams and a mid-stream pier in 1963.

SMITH

1973	Lattice	40'	15'/ -	46'	Cummings Const.

Other name(s): Pomfret Bridge. Built using trusses salvaged from the Garfield Bridge (1870) of Hyde Park. Serves leased farm fields.

TAFTSVILLE

1836	Unique truss	189'2"	17'9"/10'7"	198'6"	Solomon Emmons

VAOT #45. Modified multiple kingpost-queenpost truss with arches added. Consists of two spans. Triangular steel knee-braces added during 1953 renovation. Distribution beams added under deck in the 1980s. Overlooks power station and dam.

UNION VILLAGE

1867	Multiple Kingpost	112'7.5"	15'3"/11'3"	122'	unknown

VAOT #7. At 112 feet, it is the longest multiple kingpost span in the state. The span was once strengthened with a "timber inclined arch ... gives little structural support" (From 1993 VAOT inspection report). Timber deck replaced and distribution beams and steel knee braces added in 1970s.

WILLARD

c. 1919*	Lattice	123'5"	15'11"/13'	123'5"	unknown

VAOT #22. Serving dead-end Mill Road, the Willard stands over a mill pond that once served a woolen mill. Maintenance was done 1953 and 1979. Distribution beams were added under deck. *Unconfirmed VAOT date.

Location	Stream	Orientation	Status Capacity	Deck Girders?
Woodstock	Ottauquechee River	NE-SW	in use / 13T	no
Woodstock	Ottauquechee River	NW-SE	in use / 8T	no
Thetford	Ompompanoosuc River	NE-SW	in use / 30T	yes
Pomfret	Barnard Brook	NE-SW	private	yes
Woodstock	Ottauquechee River	NE-SW	in use / 8T	no
Thetford	Ompompanoosuc River	E-W	in use / 4T	no
North Hartland	Ottauquechee River	NE-SW	in use / 7T	no

Build Date	Truss Type	Truss Length	Portal (W/H)	Overall Length	Builder

TOUR THIRTEEN—THE WINDSOR AREA

BALTIMORE

1870	Lattice	44'11"	15'2"/ -	53'	Leland, Allen

Moved to park and restored by Milton Graton in 1967. One of three bridges moved when the Springfield-Weathersfield dam was built. Entered on Vermont's List of Historic Sites. Stands next to Eureka School House.

BESTS

1890	Tied arch	37'5"	13'10"/12'6"	45'4"	A.W. Swallows

VAOT #34. Other names(s): Swallow's Bridge, Rush Meadows Bridge. One of only three tied-arch truss bridges surviving in Vermont. Span sheltered by free-standing post-and-beam shed. Refurbished in 1991. Two new laminates subsequently added to arches.

BOWERS

c. 1919*	Tied Arch	45'4"	13'10"/11'5"	45'4"	unknown

VAOT #33. Other name(s): Brownsville Bridge, Westgate Bridge. One of only three tied-arch truss bridges surviving in Vermont. Span sheltered by free-standing post-and-beam shed. *Per VAOT covered bridge inspection report.

DOWNERS

c. 1840	Lattice	121'4.5"	15'7"/12'	131'	James Tasker

VAOT #66. Other name(s): Upper Falls Bridge. Serves Upper Falls Road. Gable end finished with splined planks and with partial cornice returns. Remarkable stone abutments. Bridge repaired in 1975–76 by Milton Graton and Sons.

MARTIN'S MILL

1881	Lattice	136'11"	16'/12'4"	147'	James Tasker

VAOT #23. Bridge site adjacent to ruins of a saw mill and lumber yard. Repairs were made in 1979, adding distribution beams under deck stringers, steel diagonal sway braces, and rods between the upper chords for lateral reinforcement.

SALMOND

1880	Multiple Kingpost	53'1.5"	16'6"/11'8"	61'	James Tasker

VAOT #83. Serves Henry Gould Road, stands next to Murray Memorial Park. One of three bridges moved when the Springfield-Weathersfield dam was built. Moved to park from the Amsden town works and restored by Milton Graton.

SMITH

1973	Lattice	39'10"	14'8"/ -	45'10"	Cummings Const.

Other name(s): Ascutney Bridge. This privately owned bridge was built using trusses salvaged from the Garfield Bridge (1870) of Hyde Park. The bridge roadway is reinforced with steel beams.

Location	Stream	Orientation	Status Capacity	Deck Girders?
Springfield	unnamed brook	NW-SE	foot bridge	no
West Windsor	Mill Brook	N-S	in use / 6T	no
West Windsor	Mill Brook	N-S	in use / 6T	no
Weathersfield	Black River	NE-SW	in use / 3T	no
Hartland	Lulls Brook	SW-NE	in use / 3T	no
Weathersfield	unnamed brook	N-S	in use / 6T	no
West Windsor	Mill Brook	N-S	private	yes

Build Date	Truss Type	Truss Length	Portal (W/H)	Overall Length	Builder

STOUGHTON

1880	Multiple Kingpost	46'2"	11'6"/ -	50'	James Tasker

Other name(s): Titcomb Bridge. Former Stoughton Bridge moved from the Black River to Titcomb's meadow by Milton Graton in 1967 and restored. One of three bridges moved when the Springfield-Weathersfield dam was built.

TOUR FOURTEEN—ROCKINGHAM TO GRAFTON

BARTONSVILLE

1870	Lattice	150'10"	15'2"/11'1"	161'2"	Sanford Granger

VAOT #44. Other name(s): Williams River Bridge. Serves Lower Bartonsville Road. Distribution beams tie-rodded under wooden deck. Steel rods were installed under deck to increase lateral strength.

HALL

1982	Lattice	120'1"	12'/14'3"	122'	Milton Graton

VAOT # 43. Other name(s): Barber Park Bridge, Osgood Bridge. Serves Hall Bridge Road. Original bridge was built at this site by Sanford Granger, in 1867. The bridge collapsed in 1980 and an authentic replica was built in 1982 using traditional materials and methods. Once known for external sway braces, no longer there. Planks used in lattices and chords are 4" thick vs. the usual 3" planks. The 12' portals, like those of the original, are the narrowest of all of the lattice bridges.

KIDDER HILL

1870	Kingpost	67'7"	12'5"/67'7"	67'7"	unknown

VAOT #41. Longest existing kingpost span. Iron rods serve as kingposts (king rods). Span is skewed. Only kingpost bridge in state with sway braces. Trusses in poor condition. Reconstructed in 1995 with 1' x 5' laminated wood (Glu-lam) beams supporting deck.

MACMILLAN

1967	Stringer	60'5.5"	12'6"/60'5.5"	62'5.5"	S. MacMillan

Simulated kingpost bridge serves foot and bicycle traffic. Timbers used are salvaged from an earlier structure. Features iron rods for kingposts. Stands next to a cheese retailer.

VICTORIAN VILLAGE

1967	Kingpost	43'9"	14'/ -	45'5"	Stratton*

*Assembled the Victorian Bridge from materials from the Depot Bridge of Townshend, which was built as a queenpost bridge in 1872 by H. Chamberlin. Now part of the Vermont Country Store complex.

WORRALL

1868	Lattice	82'8"	14'1"/11'10"	94'4"	Sanford Granger

VAOT #40. Features 20-foot wooden ramp in roadway at NE end. The truss, unique in that it lacks secondary chords, is reinforced by six pairs of 7' x 6" posts. Distribution beams tie-bolted under deck. The whole structure is stabilized with steel cables and iron rods.

Location	Stream	Orientation	Status Capacity	Deck Girders?
Weathersfield	unnamed brook	W-E	private	no
Rockingham	Williams River	NE-SW	in use / 3T	no
Rockingham	Saxtons River	N-S	in use / 11T	no
Grafton	Saxtons R., South Br.	E-W	in use	yes
Grafton	Saxtons R., South Br.	E-W	private	no
Rockingham	unnamed brook	NW-SE	private	no
Rockingham	Williams River	NE-SW	in use / 16T	no

Build Date	Truss Type	Truss Length	Portal (W/H)	Overall Length	Builder

TOUR FIFTEEN—THE DEEP SOUTH

CREAMERY

| 1879 | Lattice | 80'1" | 16'2"/11'9" | 84' | A.W. Wright |

VAOT #30. Slate roof and external walkway added in 1917. Two guy cables from west side provide lateral support.

GREEN RIVER

| 1870 | Lattice | 104'6" | 14'10"/ - | 104'6" | Marcus Worden |

Features U.S. Postal Service mail box inside, only one in state. Also has several private mail boxes. The town gin-pole is stored atop the upper lateral bracing. Bridge stands below dam and mill pond. *Posted limit.

SCOTT

| 1870 | Lattice & Kingpost | 277'2.5" | 16'3.5"/ - | 279'2.5" | H. Chamberlin |

Features one lattice span and two kingpost spans. The lattice section, 166'3.5" long, lacks upper secondary chords, and features external sway braces. Laminated arch and center pier added later. Arches anchored to abutment and pier below main stringers. Note spectacularly failed arch. Is supposedly the second longest wooden bridge in Vermont after the West Dummerston Bridge, but, as measured, the Scott is the longest wooden bridge under one roof in the state.

WEST DUMMERSTON

| 1872 | Lattice | 267'4" | 18'1"/ - | 272'10" | Caleb B. Lamson |

VAOT #35. Is supposedly the longest wooden bridge within the state. Other contender: Scott Bridge. The West Dummerston is the longest in two spans. *Closed in 1995 by the VAOT, slated for construction in 1996.

WILLIAMSVILLE

| c. 1870* | Lattice | 115'6" | 17'/11'10" | 117'6" | Unknown |

VAOT #17. Roof gable-end overhang on west end only: 2 feet, with 7 small corbels. Previously strengthened with steel beams, removed and replaced with wood beams in 1980. *N.G. Templeton estimates 1860.

Location	Stream	Orientation	Status Capacity	Deck Girders?
Brattleboro	Whetstone Brook	E-W	in use / 3T	no
Guilford	Green River	E-W	in use / 8T*	no
Townshend	West River	E-W	retired	no
Dummerston	West River	E-W	status* / -	no
Newfane	Rock River	E-W	in use / 3T	no

A SUMMARY OF VERMONT'S COVERED BRIDGES

Vermont's Covered Bridges By Truss

Truss Type	Count	Reinforced
Queenpost	21	10
Kingpost	4	3
Lattice	41	6
Timber Lattice	1[1]	0
Paddleford	3	0
Burr Arch	9	2
Howe	4[2]	0
Multiple Kingpost	8	0
Multiple Kingpost Half High	2	2
Tied Arch	3	0
Town-Pratt	1[3]	1
Unique	2[4]	1
Haupt	1	1
Pratt	1	0
Totals	101[5]	26

[1] Connecticut River Bridge at Windsor.
[2] Two are Connecticut River highway bridges.
[3] Railroad bridge.
[4] Red Bridge and Taftsville Bridge.
[5] The stringer bridge in Grafton is not counted.

Sources

For build dates, builders' names, the names of the bridges, and bridge alterations, the author used the following sources.

Publications

Branthoover, W.R. et al. *Montgomery, Vermont: The History of a Town*. Burlington, VT: The Montgomery Historical Society, 1976.

Graton, Milton S. *The Last of the Covered Bridge Builders*. Plymouth, NH: Clifford-Nicol, Inc., 1978.

Hagerman, Robert L. *Covered Bridges of Lamoille County*. Essex Junction, VT: Essex Publishing Co., 1972.

Morse, Victor. *Windham County's Famous Bridges*. Brattleboro, VT: The Book Cellar, 1960.

Northfield in the Bicentennial Year 1976. Northfield, VT: Northfield Bicentennial Committee, 1976.

Spargo, John. *Covered Wooden Bridges of Bennington and Vicinity*. Bennington, VT: Bennington Historical Museum and Art Gallery, 1953.

Templeton, Neal G. *Vermont's Covered Bridges.* Community Service Publication, First Vermont Bank, twelve-page pamphlet.

BRIDGE INVENTORIES

Vermont Agency of Transportation Covered Bridge Study, which includes structural evaluations of the seventy-five bridges in the study.

National Register of Historic Places inventory. Nomination forms for each bridge were prepared by the Vermont Division of Historic Sites. The nomination forms cite as sources:

Allen, Richard Sanders. *Covered Bridges of the Northeast.* Brattleboro, VT: Stephen Greene Press, 1957.

Congdon, Herbert Wheaton. *The Covered Bridge.* Middlebury, VT: Vermont Books, 1970.

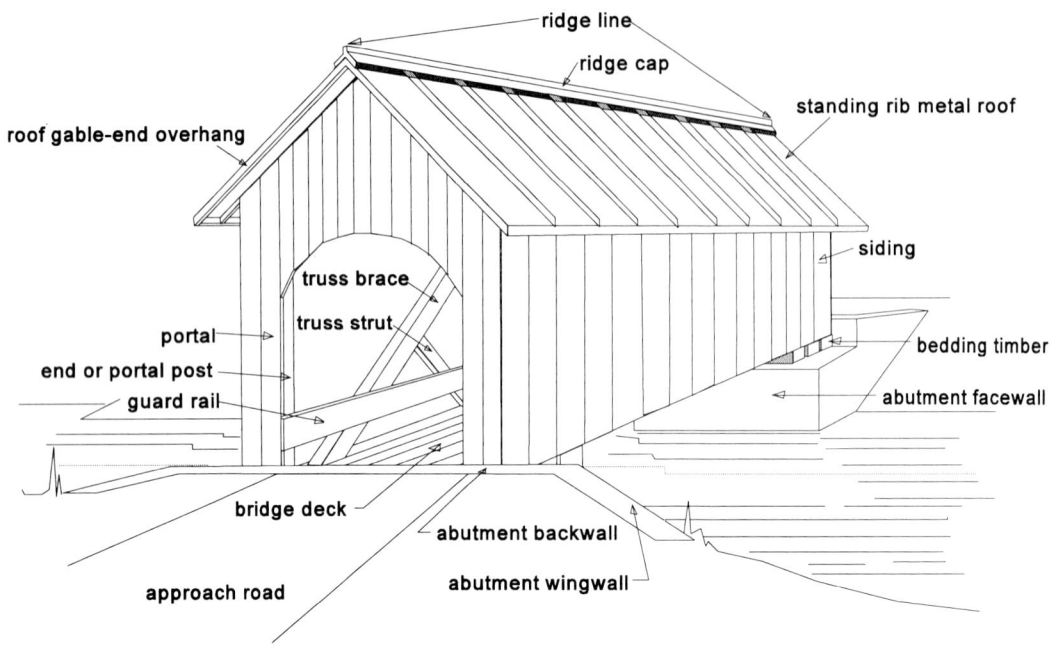

These drawings show a perspective view of a typical covered bridge and a side view of a queenpost truss.

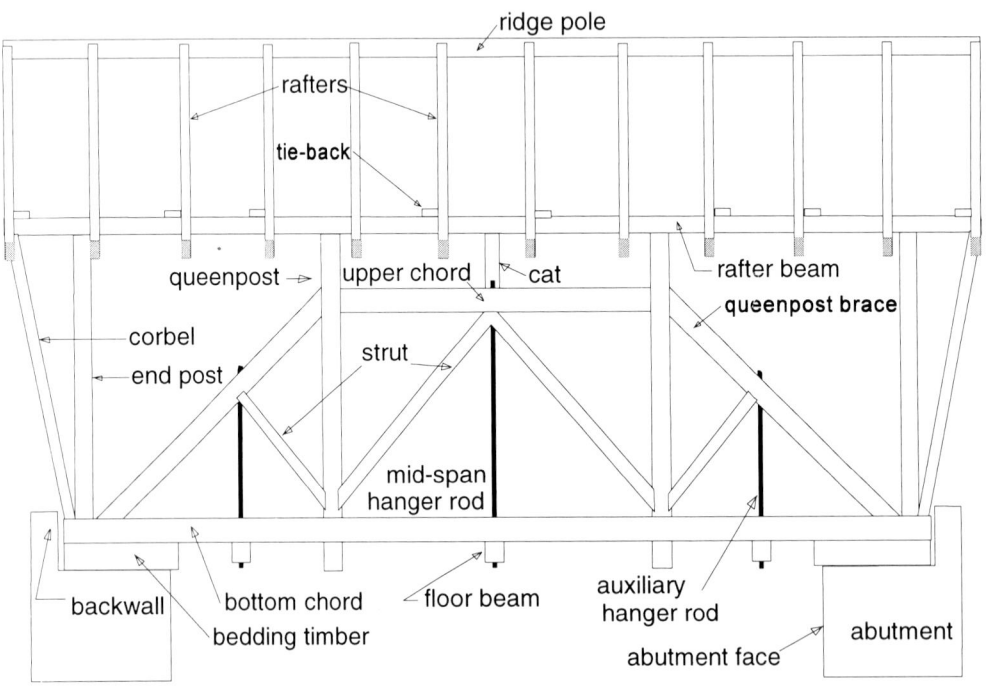

A COVERED BRIDGE GLOSSARY

Appendix B

Abutment. The abutments support the bridge at each shore of a stream. An abutment consists of a **facewall**, **backwall**, and **wingwalls**. The **facewall** is the broad side of the abutment facing the stream. The top of the abutment, above the facewall, supports the **bedding timbers**, upon which the main chords of the bridge are set. The **backwall** is constructed atop the abutment and behind the chord bedding area to serve as a retainer wall for the road bed. The **wingwalls** extend back from the **facewall** to stabilize the side slopes of the approach roadway embankment. The abutments are built upon bedrock, stone footing, or driven wooden pilings. A **backfill** of loose stone is laid behind the abutments to provide expansion space to keep frost or ice from pushing the abutment walls. (See figure on page 230)

Adz. Or **Adze**. A tool used for finishing the surface of a log squared with a **broad axe**, or for making a joint. It consists of an arched blade set at a right angle on a long ash or hickory handle. The adz-man chops at the top side of a log positioned between his feet. The tool is notoriously unforgiving of careless users—an old timer claimed he could always recognize the adz-man in a crowd by his limp. (see figure)

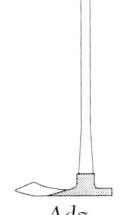
Adz

Auger. A large drill-bit fitted with a T-handle or driven by a brace or a **boring machine**. Used for drilling holes in timber.

Angle block. A triangular block of wood or iron placed at the junction of a post, brace, or arch, serving as a seat, as in the Howe truss.

Arch. In wooden bridges, a curved timber, or arrangement of timbers, forming an arc used to support or brace a **span**. In Vermont's covered bridges, arches are most commonly used

231

together with a truss in an auxiliary or supplemental role—a true arch bridge is supported only by the arch.

Timber arch. A series of timbers bolted to a truss, ending either at the faces of the abutments or piers, or at the ends of the truss bottom chords. The Burr arch truss is an example.

Laminated arch. An assembly of planks bolted together to form an arc. The ends of the member boards are staggered to maintain arch integrity. See **Inverted arch**.

Bearing blocks. Timber components used to shim between two bridge components (e.g. between a **bolster beam** and the lower truss **chord**.)

Beetle. (Pronounced biddle.) A mallet used to drive treenails into holes drilled into Town-lattice planks or pegs into mortise and tenon joints. (see figure)

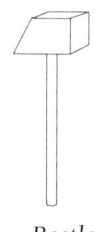

Beetle

Bed timbers. Timber components typically located between the top of an abutment or pier and the underside of the truss bottom chord. Bed timbers serve as sacrificial components—they are easily replaced when deteriorated from rot, thus protecting the truss components from similar deterioration. Also known as bedding timbers.

Bent. 1. A type of pier consisting of two or more column-like members connected at their tops. Some bents use **piles** as the column members, while others use planks or squared timbers. 2. Temporary "A" shaped frames braced with diagonal timbers designed to support a platform across their tops. Originally used to support temporary scaffolding while building a bridge, bents are now most often used as temporary supports under a damaged bridge or while a bridge is under repair. See **False work**. (see figure)

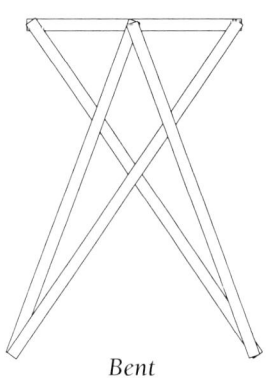

Bent

Bolster beams. Timbers used to reinforce, or bolster, the bottom chords of a bridge truss. They are placed atop the **bed timbers** and directly under the bottom **chords**, where they project behond the face of the abutment, out over the stream. Bolster beams are most commonly seen under Town-lattice trusses.

Boring machine. A drill bit holder geared like an egg beater, driven manually by two handles.

Brace. A diagonal timber in a truss that slants upward toward the mid-point of the bridge.

Bracing. A system of **tension** or **compression** timbers providing support and stability for trusses.

Broad axe. A chopping tool with a broad, sharp blade beveled on one side, the broad axe was used to square-up log timbers. The log is notched at intervals, then the chunks of

Broad axe

wood between the notches are hacked out until the log is square. A broad axe is less dangerous to the feet and ankles than an **adz**. (see figure)

Bridge deck or **Bridge roadway**. In early times the bridge deck was constructed with thick transverse floor planks supported by longitudinal joists, or stringers. The joists were supported by transverse floor beams. As years passed and loads grew heavier, the floor beams were spaced much closer together to provide more support. Sometimes, the floor beams were close enough that the joists were no longer needed, and longitudinal deck planks were fastened directly to the beams. A variation is the practice of turning two-by-six or two-by-eight planks on edge and nailing them together, butcher block style, to form a "nail-laminated" deck.

Buttress. An assembly of timbers or iron rods placed along the outside of both sides of a bridge and connected to the ends of extended floor beams. The upper end of the timber or rod is attached to the top of the truss-work. Sometimes called a **sway brace**.

Camber. A curvature provided to compensate for **dead load** deflection, bridge camber is an upward bowing of the **chords**. A chord is usually several inches higher in the middle of the span than it is at its ends. Incidentally, camber compensates for the optical illusion of a sag in a straight line. A truly sagging bridge is said to have negative camber.

Check brace. A brace designed to aid the kingposts in resisting the compressive forces transmitted by the main braces. Also called a chalk brace or kicker brace.

Chord. 1. The upper and lower longitudinal members in a truss, extending the full length of the truss and carrying the forces of tension and compression away from the center of the span. 2. The top (upper chord) or bottom (lower chord) member or members of a bridge truss. A chord may be a single piece of timber or a series of joined pieces. Town-lattice trusses typically contain two levels of top and bottom chords; hence, there may be upper and lower top chords and upper and lower bottom chords.

Clear span. The span of a bridge measured from the facewalls of the abutments.

Compression member. An engineering term that describes a timber or other truss member that is subjected to squeezing or pushing. Also see **tension member**.

Corbel. A timber projecting from the end post of a portal to support an overhanging gable-end. Corbels are often used as decorations.

Cornice. A horizontal molding installed under a roof-line to make a finished intersection of roof and sidewalls.

Cornice return. When a cornice molding is wrapped around the gable-end of a building or covered bridge, it is called a cornice return. When the molding is stopped just past the eaves on the gable-end, it is called a **partial cornice return**. (See figure on page 237)

Counter brace. A diagonal timber in a truss that slants in the opposite direction from the **brace**. It is usually a tension member.

Crib. Assembled from timbers or logs stacked to support falsework, cribs are used when putting a bridge across a stream or to support jacks when raising a bridge for repairs. When the timbers are spiked together and the assembly filled with stone, the crib can serve as a semi-permanent pier. (see figure)

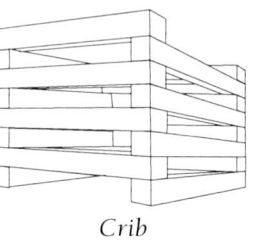

Crib

Cross brace. Cross member spanning from truss to truss at roof level or under the floor.

Cross-x bridge. A colloquial name for a bridge built with a plank-lattice truss. Also known as an x-bridge.

Dead load. The static load imposed by the weight of materials that make up the bridge structure itself.

Distribution beams. Longitudinal timber components aligned below, and supported by, the floor beams of the structure. These beams have been installed under existing deck systems in recent years and are intended to force the participation of several floor beams in supporting axle loads of vehicles. Computer analysis and load tests have shown them to be of negligible structural benefit, but current thought is to leave them in place unless extensive work is done to the deck. In that case, distribution beams that are removed for work on the floor system are not replaced.

Double-barreled bridge. A colloquial name for a covered bridge with two roadways. There is usually a third truss between the lanes. The Pulp Mill Bridge in Middlebury and the Museum Bridge in Shelburne are examples.

Draw knife. A wood shaving tool consisting of a blade beveled on one side and positioned between two handles. The blade can be wide and flat, or formed for spoke or pin shaping. The user, holding the tool in both hands, "draws" the tool toward himself, with the grain of the wood. Timber bridge builders use it for shaping treenails and pegs for securing mortise and tenon joints. (The treenails used in plank-lattice bridges were invariably turned on a lathe for accuracy of fit.) (See figure)

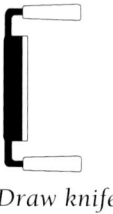

Draw knife

Falsework. Consists of a temporary shore-to-shore platform supported by **cribs** or **bents**. Falsework is used to support a bridge under construction or major repair. See **scaffolding**.

Floor. The road bed or deck of a bridge.

Floor beams. The beams laid between the bottom chords of the bridge trusses. The floor beams support the **joists** (or stringers) upon which the **transverse** floor planks are laid.

Freshet. A flood caused by heavy rains or melting snow.

Froe. A wide shallow wedge with a handle set at a right angle on the end of the blade. The tool is used to split shingles from a block of pine or cedar. The user holds the handle upright and strikes the top of the blade with a maul.

Fulling mill. A mill that treats cloth, cleaning and thickening (or felting) cloth by beating and washing it. Fulling mills were once common in Vermont in areas where sheep were raised for wool.

Gin pole. A pole or timber erected in a leaning position, restrained by two cables secured to the top of the pole and anchored to positions to the rear of the pole. Pulleys and ropes hung from the top of the gin pole are used for lifting heavy objects, including the trusses of a bridge.

Gusset. A plate serving to connect the members of a joint and hold them in correct alignment and position. A gusset serves the same function as that of a **ship's knee**.

Hanger rod. See **suspension rod**

Inferior brace. A timber or beam slanting upward from the face of an abutment or pier to the underside of the bridge, usually lending additional support to the main bridge **chords**. It's so named because the brace occupies a position below that of all other braces and beams. Also called arch bracing.

Inverted arch. Deformation of an arch caused by over-loading the bridge or by the incorrect adjustment of suspension rods in a laminated arch bridge. Over-tightening or tightening the suspension rods in an incorrect order can result in an inverted arch. The problem can be seen in the Scott Bridge.

Joist. Timbers laid longitudinally on the floor beams of the bridge. The floor planking is laid over these. Also known as a **stringer**.

Kingpost. In a kingpost truss, the vertical wooden post hung from the apex of the main diagonal braces. In a multiple-kingpost truss, the vertical member is paired with one or two diagonal braces.

King-rod. An iron rod substituted for, or augmenting, a **kingpost**.

Key. A wedge inserted into a mortise and tenon joint to tighten it.

Knee brace. 1. A short timber bent at a right angle used inside a covered bridge between a truss and upper lateral bracing to increase rigidity. 2. Transverse timber component connecting the upper portion of the truss with the transverse tie beams, usually positioned at a 45-degree angle. See **ship's knee**.

Late blooming tourist. A tourist who explores Vermont after the leaves are down and before the snow flies.

Lateral bracing. The use of timbers between the two bridge trusses, spacing them apart and creating a single unit to brace the bridge against **transverse** forces.

Longitudinal. The direction parallel to the bridge span.

Mortise. A rectangular cavity cut into a timber to receive a tenon. See **tenon**. (see figure)

Panel. A rectangular section of truss between two vertical members, bounded top and bottom by upper and lower chords, including all struts and braces. Some well defined panel systems are the Long, Howe, and Pratt trusses, as well as the basic multiple kingpost truss.

Patented truss. Those truss types for which United States patents have been granted. Examples are the Burr, Town, Long, Howe, and Haupt trusses. The kingpost, queenpost, tied-arch, and Paddleford trusses were not patented.

Pediment. The triangular space formed by the sloping roof at a gable end. (see figure on page 237)

Penstock. A channel or pipe carrying water to a water wheel or turbine.

Pier. A wood or masonry bridge support built in the stream bed between the abutments.

Pilaster. A false pillar, usually a decorated plank used to adorn a bridge portal. The plank is decorated with grooves to imitate a fluted column or trimmed with moldings to suggest the capital of a Doric column. (see figure on page 237)

Pile. A large, straight pole, most often a debarked log, driven into soft, wet, or submerged sites to provide a secure foundation for a bridge abutment or pier.

Pit saw. A two-man saw used to rip-cut a log or timber. The log is placed over a pit, with a man on top and another in the pit. The man in the pit is usually an apprentice or the man "lowest in the food chain," as one bridge engineer puts it. "The man on top provides the return stroke of the saw, the man on the bottom uses his body weight to provide the cutting stroke, earning a shower of saw-dust for his efforts. This is where the phrase 'in the pits' comes from." See **rip**.

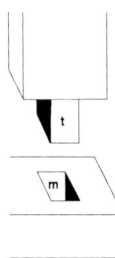

m-mortise
t-tenon

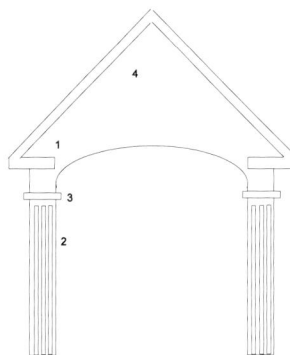

Bridge portal c. 1840
(1) partial cornice return
(2) pilaster
(3) pilaster molding
(4) pediment

Plate. The uppermost horizontal member supporting the lower ends of the roof rafters. Also known as a **rafter beam**.

Pointing. The practice of forcing mortar into the joints in the stonework of a bridge abutment.

Portal. General term for the entrance of a covered bridge. (see figure)

Post. The upright or vertical timber in a bridge truss.

Queenpost. One of two vertical posts in a queenpost truss suspended from the apex of a diagonal brace and the upper chord.

Queen-rod. An iron rod substituted for, or augmenting, a **queenpost**.

Rafter beam. See **plate**.

Rafters. The rows of beams arranged in the form of inverted V's that support the roof planks of a covered bridge.

Rip. To saw a plank or timber lengthwise.

Roadway. See **bridge roadway**.

Rod. Iron or steel rods used as vertical tension members in arch truss bridges to support the lower chords. Bridge members could be tightened and camber maintained by adjusting nuts against washers on the ends of the rods. The Howe truss features iron rods and **turnbuckles**. Also see **king-rod** and **queen-rod**.

Running planks. Longitudinal planking placed over a bridge deck to provide an easily replaceable wearing surface. They are usually found over bridge decks constructed of transverse planks.

Sag. A permanent downward deflection of trusses at the middle of the span, also known as negative camber. Sag is not good—it is evidence that the truss is failing.

Scaffolding. A temporary wooden platform built to support the erection or repair of a bridge. See **falsework**, **bent**, and **crib**.

Scarf joint. A joint in which the ends of beams are cut so that they overlap and join firmly. Scarf joints are used to splice chords or stringers end-to-end. See **tension splice**.

Secondary chord(s). The chord or chords between the upper and lower chords in a plank lattice truss.

Shear. A force causing two parts or pieces to slide on each other in opposite directions.

Shear key. A spline, or piece of hardwood, usually oak, fitted into a slot or groove between two parts to resist shear forces.

Shelter panel. The first panels at the portals of a covered bridge are boarded over on the inside to protect the timbers from moisture. The practice is said to have begun after the advent of automobile traffic. Also called false doors or splash panels.

A COVERED BRIDGE GLOSSARY

Ship's knee. A timber cut from a tree branch or root found naturally bent at a right angle. It is used to reinforce the joint between the bridge trusses and the upper lateral bracing. Sometimes called **knee braces**, they can be seen in the Big Eddy and Fisher bridges.

Sightly. Colloquial Vermont expression describing a pretty or handsome object, such as a particularly appealing covered bridge. A traveling Vermonter, upon seeing Pikes Peak, described the scene as "sightly."

Sister. Additional Town Lattice web member inserted adjacent to a damaged or deteriorated web member to increase the strength of the truss without replacing the existing member.

Skew-back. A step or notch in the face of an abutment to receive the end of a chord or an arch.

Skewed bridge. A bridge built to cross a stream at other than 90 degrees.

Snowing the bridge. The act of covering a bridge floor with snow to allow sleighs to be drawn through.

Span. The length of a bridge between **abutments** or **piers**. **Clear span** is the distance measured from the face of one abutment to the face of the other. The length of a covered bridge is usually calculated using the truss length—the distance between the truss end posts—regardless of how far the truss may overreach the actual abutment.

Splice. A joint or the act of joining timbers end-to-end. See **tension splice**.

Splined boarding. A treatment used to fasten the long edges of sheathing boards, usually on a gable end. The long edges are grooved and a strip of wood called a spline is fitted into the grooves to serve as a key. When skillfully done, the seams between the splined planks are nearly invisible. The last example of the practice on a covered bridge in Vermont can be seen on Downers Bridge in Weathersfield.

Stringer. See **joist**.

Strut. A supporting piece in a truss that acts in support of a brace or a beam.

Suspension rods. Iron rods used to suspend a bridge deck system from an arch.

Sway brace. See **buttress**.

Tenon. A tongue shaped at the end of a timber to fit into a mortise to form a joint between a post and a beam. See **mortise**. (see figure on page 236)

Tension member. An engineering term. Any timber or rod of a truss which is subjected to pull, or stretch. See **compression member**.

Tension splice. A zig-zag or interlocking splice in a tension member, usually a bottom chord. Designed to resist slippage caused by tension forces. The interlocking surfaces are usually strengthened with a hardwood shear key. Also known as a **scarf joint** or bolt-of-lightning splice.

Tie beam. Transverse timber component connecting the tops of the trusses. A part of the upper lateral bracing system.

Transverse. 1. A direction—from side to side, as in planking cut and installed at right angles to the length of a bridge-deck. 2. Forces, such as those from wind or water, acting upon a bridge structure from the side.

Transverse bracing. The bracing between the trusses to resist lateral forces and provide stability.

Treenails. Pronounced "trunnels." The wooden pins driven into the holes drilled into the plank members of a lattice truss to fasten them together. Treenails are also used to pin a mortised joint together. Treenails are made of hardwood, usually oak.

Truss. A framework of beams usually connected in a series of triangles, used to support a roof or a bridge. The triangular element in the truss is desirable because the triangle is inherently stable and resists deformation.

Turnbuckle. An internally threaded metal loop used to tighten iron rods or steel cables.

Wedge. See **key**.

Windbracing. The lateral bracing in the roof and under the floor designed to brace the bridge against transverse forces, such as the wind.

Glossary Sources:

Allen, Richard Sanders. *Covered Bridges of the Northeast.* Brattleboro, VT: Stephen Greene Press, 1957

Graton, Milton S. *The Last of the Covered Bridge Builders.* Plymouth, NH: Clifford-Nicol, Inc., 1978

Lewandoski, Jan. *Wood Truss Highway Bridges in North America: Repair and Strengthening.* Proceedings of the Fifth International Conference on Structural Faults and Repair held at the University of Edinburgh, Scotland, June 29, 1993.

Ritter, Michael A. *Timber Bridges, Design, Construction, Inspection, and Maintenance.* Washington, DC: United States Department of Agriculture, 1990

Vermont Agency of Transportation Covered Bridge Study. Prepared for the State of Vermont Agency of Transportation by McFarland-Johnson, Inc., Binghamton, NY, 1995.

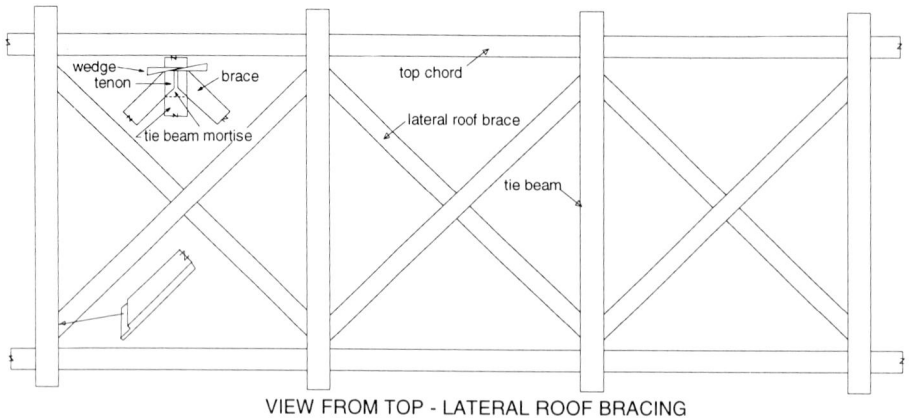

VIEW FROM TOP - LATERAL ROOF BRACING

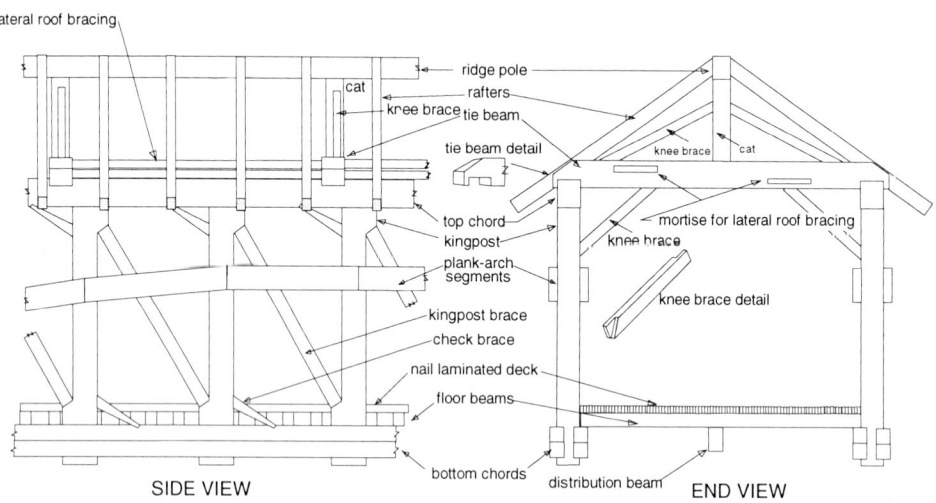

SIDE VIEW   END VIEW

The above drawings show construction and bracing details of a multiple-kingpost bridge with a Burr arch. The information is taken from the Poland, Grist Mill, and original Gates Farm bridges. The drawings below show chord-splice assemblies for top and bottom chords.

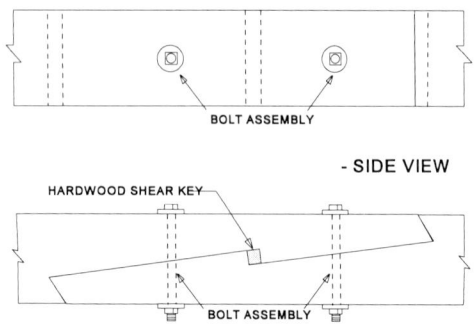

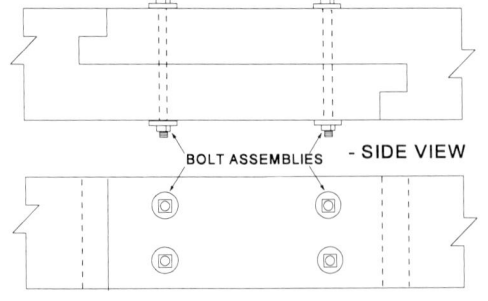

The BRIDGE TRUSS

Appendix C

The Bridge Truss

A truss, according to most dictionaries, is a framework of beams, usually connected in a series of triangles, used to support a roof or a bridge. The triangular element in the truss is desirable because the triangle is inherently stable and resists deformation. This definition is perfectly adequate for the discussion that follows.[1]

A clever carpenter, with the shared experience of his peers, could put together a successful truss if the timbers he used were stout enough. Today it is called the empirical method, one based on practical experience, observation, and experimentation. In the early years, when the republic was in dire need of bridges, people in the countryside, the farmer-carpenters who participated in community barn raisings, were not afraid of attempting to build a bridge. Using the same principles of construction as used in the barns, churches, and town halls, they built the stringer, the kingpost, and the queenpost truss bridges ignorant of the finer points of engineering.

Bridge builders continually experimented with trusses so that their bridges could span longer distances and carry heavier loads. The work of Theodore Burr, Peter Paddleford, and Ithiel Town reflects early progress in this area. As the country industrialized and placed ever greater demands on highway and

---

[1] A basic engineering text might define the truss as a jointed structure having an open web construction so arranged that the frame is divided into a series of combined triangles with each frame member primarily stressed along its own axis. This says essentially the same thing as the definition above, except that it further states that the truss is constructed to balance stress forces acting in a single plane.

railroad bridges, however, more scientific approaches had to be used to make economical use of stronger, lighter—and more expensive—construction materials. Beginning in the 1830s, with the railroad boom, trained engineers such as Stephen Long, Willis Pratt, Herman Haupt, and William Howe took over the job of designing new trusses, but it was not until Howe patented his bridge in 1840 that a design included a complete mathematical analysis of truss stresses.

Long before the engineers came on the scene, the old master bridge builders had developed the rules considered good timber engineering practice today. They followed principles such as: keep the wood dry and ventilated; choose, cut, and dry your timber carefully; beware of sloping grain; use wood shrinkage and expansion to your advantage; use long continuous members when possible; be careful how you join two pieces of wood, and where you put the joint. These practices are responsible for the fact that so many of these bridges are still standing today after one-hundred-plus years.

"You would be surprised what punishment a wooden bridge can take," marvels wooden bridge builder and restorer Jan Lewandoski. "They may sag or a floor beam may break, but wooden bridges very rarely fail catastrophically. One good thing about them is that they give you lots of warning—the trusses will creak, distort, make horrible noises, and scare you to death years before they fail, as opposed to steel bridges, which have a tendency to just snap."

How Do The Trusses Work?

Simply stated, in a plank-lattice truss for instance, the upper chords are in compression, carrying the weight of the bridge downward along the lattice members sloping toward the abutments. The lattice members that slant the other way, toward the center of the bridge, are in tension, as are the lower chords, stretched by the weight of the bridge and its load. Also, because the lattice truss is pinned where the members cross, there is interaction between the forces of compression and tension at these points too! If the bridge uses mid-stream piers the roles of the lattice members reverse at the piers, each span really being a separate bridge.

The magnitude of the chord and lattice stresses vary along the length of the bridge. Chord stresses build incrementally from the abutments to a maximum at midspan, while the stresses carried by the lattice members increase from a minimum at midspan and are maximum at the abutments.

When looking at the following diagrams, keep in mind that, while the members perform as described, actually calculating the forces involved is a very complex exercise that lies beyond the scope of this book. Realizing this, we can proceed from here as the old bridge builders did, unconfused by the finer points of bridge engineering.

The Kingpost Truss

The kingpost truss, in its simplest form, is found in frame buildings, where there is a need to provide large spaces without columns or load-bearing walls. The most common form found today supports the roofs of small commercial buildings and the ubiquitous post-World War II ranch-style homes—those rambling houses with the low-pitched roofs and no useful attic space. These mass-produced roof trusses, found in any lumber yard, are descended from the structures used in the huge old barns and meeting houses that had cavernous interiors, constructed with post and beam and fastened with mortise and tenon.

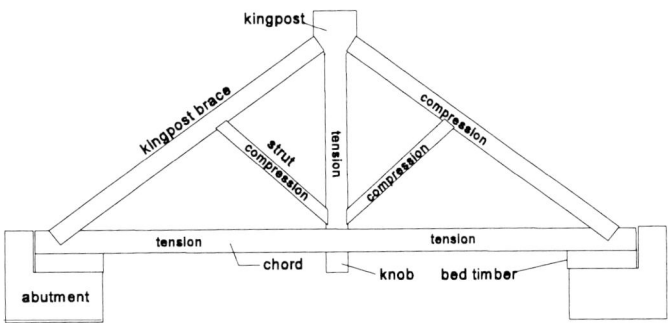

Kingpost truss

The kingpost truss was adapted to support the shorter bridge spans. It was probably first used with small open-work bridges, and it represented quite an improvement over stringer bridges, in which the trunks of trees were simply thrown over a stream.

In the kingpost truss, a timber called the kingpost is, in effect, suspended by its top end by two diagonal timbers, whose ends are braced on the ends of the chord above the abutments on each stream bank. The center of the chord is suspended from the kingpost. The weight of the bridge is transmitted to the ends of the chord through the kingpost braces. The kingpost braces are said to be in compression, squeezed between the chord-ends and the kingpost by the weight carried by the kingpost. The kingpost, being stretched by the weight of the roadway and the outward thrust of the kingpost braces, is said to be in tension. The chord is in tension as well, supported at the abutments and by the kingpost and stretched by its own weight and the weight of the roadway. As long as the truss, its joints, and its timber members are strong enough to bear the stresses of compression and tension, the bridge will stand.

The carrying capability of the kingpost truss requires that the angles between the kingpost braces and the chords not get too small, meaning that a wider stream would require a really tall kingpost. The maximum practical span for a kingpost bridge is a little over forty feet. Longer spans require a more sophisticated truss. The queenpost truss, for example.

The Queenpost Truss

The queenpost truss is actually an extended kingpost truss. The kingpost is replaced by two queenposts, braced apart at their tops by a horizontal beam and at their bases by the bridge chord. These vertical tension members may well be called queenposts because they are smaller in size than a kingpost would be for a span of equivalent length.

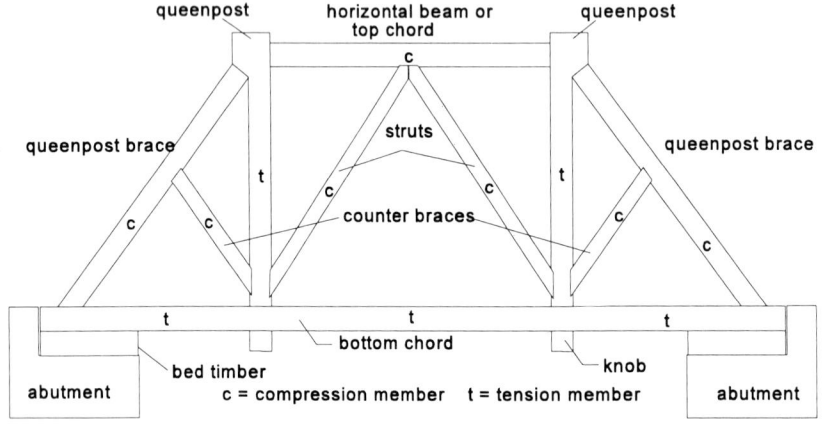

Queenpost truss

The weight of the bridge and its cargo is distributed to the ends of the chords at the abutments by the queenpost braces and the horizontal beam in the manner of a three-piece timber arch. The dimensions of the queenpost timbers are typically ten by twelve inches, and the braces and horizontal beam are nine by twelve inches. The chord construction may not be a solid timber. In some queenpost bridges, the chord is constructed from three or more planks, pinned or bolted together—the middle planks are cut out to receive the posts and braces.

The average truss-length of the queenpost bridges surviving in Vermont is approximately fifty-eight feet. The longest is the Flint Bridge, which measures eighty-eight feet. The School House Bridge in Lyndon is the shortest at forty-one feet eight inches.

The Multiple-Kingpost Truss

The multiple-kingpost truss is an excellent example of the use of a "series of combined triangles" to create a load-bearing span. The tops of the vertical members, or kingposts, are joined to the upper chord by mortise and tenon. The bases of the kingposts are joined to the lower chord by an interlocking notch or lap. (see the diagram illustrating the truss-to-bottom-chord connections on page 247.) The braces are set into steps cut into the kingposts without mortise, tenon, or tongue. The assembly of upper and lower chords, kingposts, and braces form a rigid truss that resists deformation. The weight of the bridge and its users tends to settle the truss, compressing the braces and upper chord while stretching the lower chord and the kingposts supporting it.

Vermont's multiple-kingpost bridges average sixty-seven feet in length. They range in length from James Tasker's forty-six-foot Stoughton Bridge to the Union Village Bridge, which stretches a whopping 112 feet. The kingposts typically measure six by eight inches cut from a six-by-sixteen-inch timber to provide steps for the braces. The braces typically measure four by six inches.

Multiple-kingpost truss

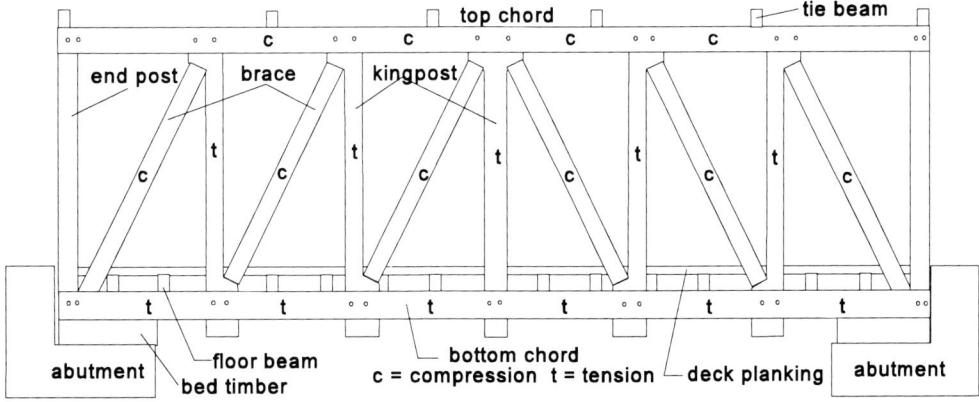

The Burr-Arch Truss

Some of the early builders of wooden bridges experimented constantly with new trusses. The goal was to find a truss that would allow spans to cross ever-wider streams with fewer piers. Long clear spans are important in New England, where rivers in flood are particularly unforgiving to obstructions, hurling ice, trees, and chicken coops at them, depending on the season. A pier tends to collect such flotsam, building dams on its upstream side. The burden of the dams and the water behind them can displace and destroy the bridge.

Theodore Burr built a bridge across the Hudson River in Waterford, New York, in 1804. There, he used a combination truss and segmented arch that became the basis of his 1817 patent. The patented truss consisted of parallel chords tied together by vertical posts and stiffened by crossed braces. Each timber-arch segment, lap-jointed on both ends, is fitted into the next on both sides of the posts and braces. The segments are "through-bolted" to the vertical posts, sandwiching the truss. The ends of the arches are received by the abutments below the bed timbers.

The Vermont versions differ from Burr's patent in that the ends of the arch are supported by the ends of the lower chords rather than at the faces of the abutments. Also, a multiple-kingpost truss is used rather than the cross-braced truss.

Multiple-kingpost truss with Burr arch

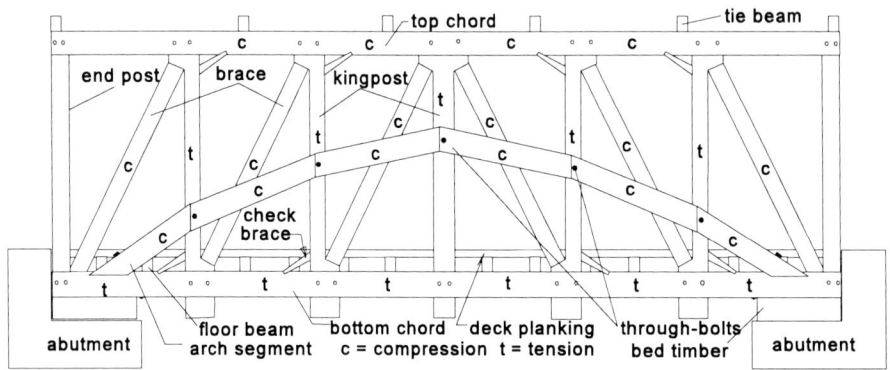

The Burr truss is not a true arch-bridge in that the parallel-chord truss system and the arch act in concert to support the bridge and its load. The multiple-kingpost truss serves the purpose of aligning the arch segments while carrying the roadbed. The arch serves to make the whole structure more rigid, and the combination is capable of longer spans than is the multiple-kingpost truss alone. The Burr-type bridge is best classified as an auxiliary arch system. In a true arch bridge, the roadway is supported only by the arch.

How the Burr truss actually works is "a thorny topic of debate," according to Gilbert Newbury, an engineer with the Vermont Agency of Transportation. Newbury, who is trained in timber engineering, planned the repair of the Burr-arch bridge on Gates Farm in Cambridge. He explains that one school of thought has it that the arches do the work, while the other believes it is the parallel chord truss.

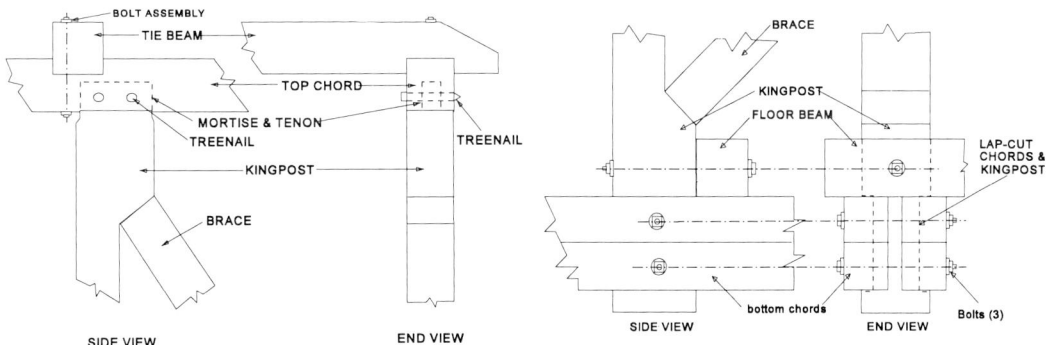

Post-to-top-chord and post-to-bottom-chord connections (Gates Bridge)

Newbury's computer analysis and field observations indicate that it is the truss that carries the majority of the bridge and roadbed weight, with the plank arch acting as a stiffening element for the truss. He bases his conclusion on three points.

First, an arch made of doubled five by twelve planks eighty feet long (the length of the Gates Farm Bridge arch) is not very strong. Second, the Burr arch lacks height, so it cannot provide the strength of a properly proportioned arch. Finally, it is characteristic with wood structures that when you join two pieces of wood together, it is the joint that breaks. In the case of the plank arch, the bolted connections joining the arch segments to the truss are not very strong if the arches were meant to carry a load. The joint used *is* adequate for stiffening the parallel chord truss, however.

The builders' intentions remain unclear. "Did the old-timers make a mistake with the bolted connection, or did they know what they were doing?" wonders Newbury. "I have found every time I think I've caught them in a mistake, it turns out to be my error."

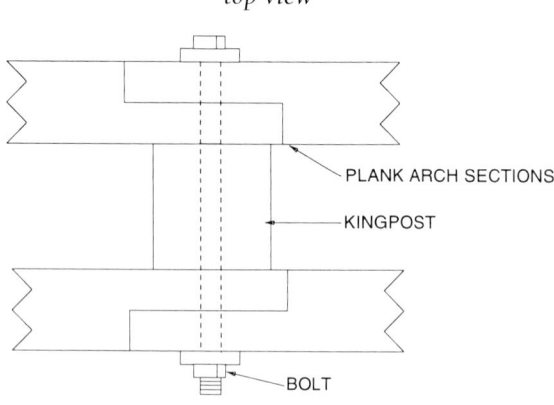

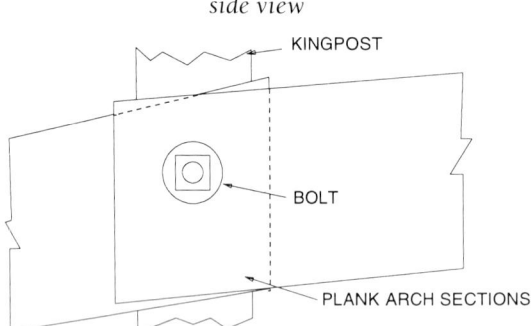

In Burr's plank arch, each arch segment is rabbeted on both ends and fitted one into the next on both sides of the kingpost and bolted into place.

Eight covered bridges based on Burr's design survive in Vermont at the time of this writing. The kingpost members of the trusses typically measure six by nine inches cut from nine-by-eighteen-inch timbers to provide steps for the braces. The braces typically measure six by nine inches. The arch segments are typically made from five-by-twelve-inch timbers.

The average length of the surviving Burr-arch bridges is 101 feet. The longest is the Museum Bridge in Shelburne at 156 feet, and the shortest is the seventy-foot Seguin or Upper Bridge in Charlotte. The 199-foot Pulp Mill Bridge in Middlebury-Weybridge is not included here since its Burr arches were replaced with a laminated-arch and iron-rod system in an attempt to cause it to function like a "true" arch bridge.

The Plank-Lattice Truss

The plank-lattice truss was first patented in 1820 and again in 1835 by architect Ithiel Town of New Haven, Connecticut. Mr. Town aggressively promoted the use of his invention, sending agents all over New England. If a builder chose to use the design, Town's agent would license him for a one-dollar royalty per foot of bridge. If a builder was found using the design without permission, the royalty fee was doubled. Town made his fortune not by building bridges himself, but by selling the rights to use his design.

The timber bridges were fastened together with mortise and tenon, tediously drilled, chiseled, and—finally—pegged. The lattice bridge is constructed entirely of planks instead of the heavy timbers used in the queenpost and kingpost trusses. The planks, usually of spruce or hemlock, are simply fastened with hardwood pegs called treenails (pronounced trunnels). A lot of treenails are needed, however—it was noted by a bridge carpenter that more than 2,500 holes must be drilled to receive nearly a thousand treenails in assembling a one-hundred-foot lattice highway bridge.

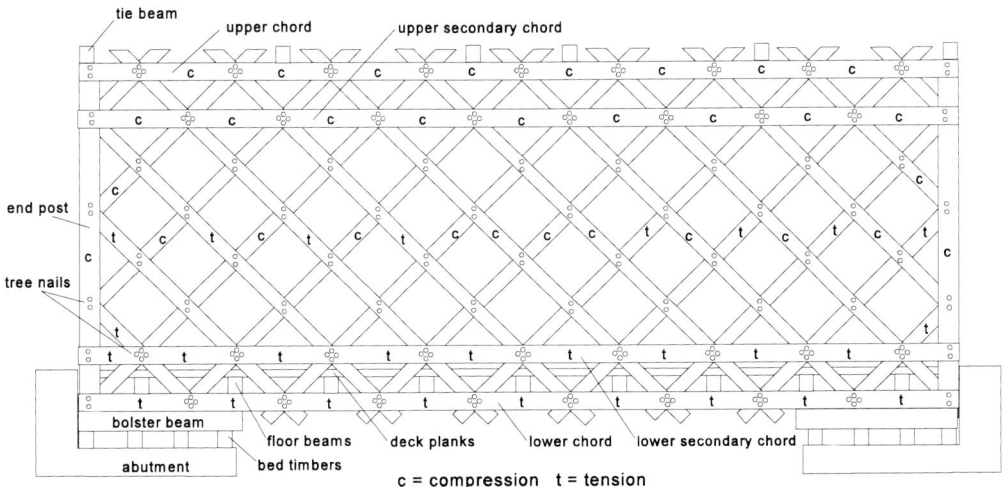

Plank-lattice truss (in the manner of Ithiel Town's patent)

The treenails are used to pin the chord planks together on both sides of the lattice and to pin the lattice planks together where they cross. The chords usually consist of four layers of planks—two layers on each side of the lattice. A strong chord requires that the planks should not be less than thirty feet long and that the plank pairs should overlap one another by half their lengths. The chord-plank pairs are pinned together at a lattice-plank crossing with three or four treenails.

The plank-lattice design calls for four chords on each truss—the lower chord and secondary lower chord and the upper chord and secondary upper chord. The bridges built with the four pairs of chords have held up well over the years against wind, water, and heavy loads. Only three of Vermont's plank-lattice bridges built without upper secondary chords survive—the School House Bridge in North Troy, the Scott Bridge in Townshend, and the Worrall Bridge in Rockingham. All three have required additional bracing.

Except for where the chords cross, the lattice-plank crossings are secured with two treenails. The School House Bridge is the exception—it uses only one treenail at a lattice crossing. That goes against everything you expect the crossing of the lattice to do, according to Jan Lewandoski.

"The pinning of the lattice truss is very important," said Lewandoski. "The bottom chords keep it from pulling apart in tension. The top chords are resisting compression, the squeezing together of the bridge when the load comes on it. Imagine if you had a lattice truss and you tried to squish it down, flex it. The lattice would act like a children's restraint gate that folds into a new shape—if there is but one pin at the crossings. Two pins keep the joint from rotating and the lattice, together with the chords, resists flexure of the bridge."

Gilbert Newbury related his experience with treenails when he restored the Silk Bridge in Bennington. Some of the other restorations have had to resort to glue to keep the pegs in the holes. "Any time you have to replace a lattice member, it automatically means you have to have new pegs. The new member should be installed green, which means the plank and the holes drilled in it will shrink as the plank dries. We had the treenail dowels turned from very dry white oak to the exact diameter of the holes. The treenails will not shrink—they can only swell as they take on moisture. Greased with beeswax, it takes about twenty blows with a sledge hammer to get them in, but once they are in, moisture transfer locks them in place. It's another example of where the old method works very well."

"When you consider the plank lattice truss," says Paul Ide, framer and joiner of bridges and barns, "the genius of it is in its simplicity and what it can do. It's obvious that it's a good truss to use wood in because wood is not predictable. Any one piece can be different by quite a bit! If you have hundreds of junctions like the plank lattice does, it doesn't matter if some pieces are weaker than others. In the queenpost truss, on the other hand, it matters a lot.

"The main thing about the plank lattice was that it was the cheapest and easiest to build. Any carpenter would be able to build it quickly, without large dimension timber. You didn't need anything thicker than three inches for your chords, and you generally did not need a plank longer than thirty-two feet."

The usual angle in the lattices constructed in Vermont is 45 degrees, with some as steep as 55 degrees. The average length of Vermont's surviving lattice truss bridges is 105 feet. Plank sizes are typically three by eleven inches with the lattice spaced at three-foot intervals. The longest lattice bridge is the West Dummerston Bridge, which stretches 267 feet in two spans. The longest single span for a bridge still in use without a re-engineered roadway belongs to the Bartonsville Bridge in Rockingham, at 151 feet. The forty-nine-foot eight-inch Fuller Bridge in Montgomery is the shortest lattice bridge on a public road.

It was once said that the Town plank-lattice truss could be built by the mile and sawed off to suit the site. The design proved to be very sturdy, and the lattice bridges had a lot of structural integrity. On several occasions, lattice bridges were dragged back to their abutments and reset after being swept away by a freshet. In one instance, a tumbled lattice bridge was used as it was found when an impatient farmer drove his team across the stream on an upturned side of the bridge. They made them like that in old New England—the farmers and the bridges.

The Long Truss

Stephen H. Long patented a parallel-chord truss in 1830. The new truss featured crossed wooden braces supporting vertical wooden posts. The patent was modified in 1836 and again in 1839. A forerunner of the panel type trusses used in iron bridges, it was the first American truss where mathematics was used in the design. The truss was used by some of the earlier railroads in New England.

There are no Long trusses still in existence in Vermont, but the truss is described here for its significance in the development of bridge trusses.

Jan Lewandoski has studied the Long Truss, has worked on bridges using the trusses in New York, and has built a new one in Maine. He explains that the Long is basically a multiple-kingpost truss with a counterbrace. In Long's design, the braces and counterbraces are not attached to the chords—they are set into the truss and held there by the forces of compression.

"When you bend, or deflect a bridge, all of the braces that rise from the abutment toward the center of the span get squeezed. The counterbraces, the ones that slant downward from the top chord toward the center of the span, get pulled apart. You would expect that the counterbraces would fall out.

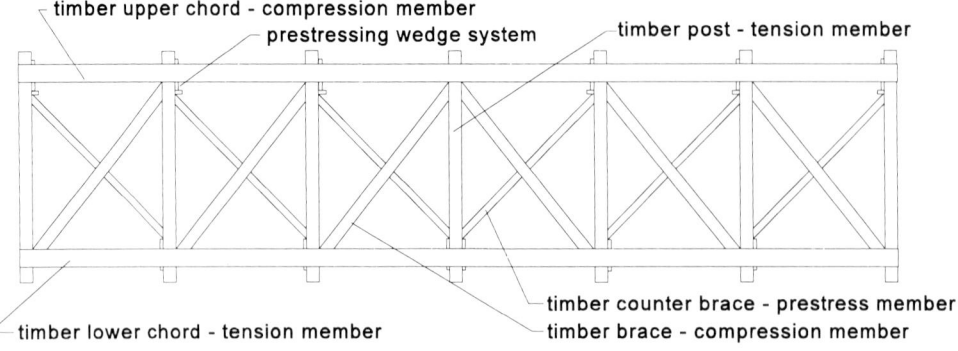

Long truss

They would, but for the special feature in Long's patent—the pre-stressing of the bridge.

"When construction is completed, and before any traffic is allowed on it, the builder overloads the bridge, causing it to deflect. Then the counterbraces are driven into place with wedges. When the load is removed from the bridge, it does not spring up—it is prestressed. When heavy loads cross a prestressed Long truss bridge, the bridge does not engage in the up and down vibration that occurs with other bridges."

A good example of the Long truss in use can be found in the covered bridge in North Blenheim, New York, completed in 1855 by Nicholas Powers of Pittsford, Vermont. The two-lane 210-foot bridge stands over the Schoharie Creek, next to Route 30, thirty-five miles southwest of Schenectady.

The Paddleford Truss

Peter Paddleford was a bridge builder from Littleton, New Hampshire. He was a user of the Long truss, employing it on the Connecticut River bridges he built at Monroe and Northumberland, New Hampshire. He created his own truss by modifying the Long truss, stiffening it with a system of interlocking counterbraces.

While the Paddleford truss design was never patented, it was widely used. It was especially popular in nearby Orleans and Caledonia counties, where Vermont's last three Paddleford bridges still stand.

The Paddleford truss resembles a multiple-kingpost truss with counterbraces. The chords are constructed from three or more planks pegged together face to face. The vertical posts are mortised into the upper chords and lapped, or notched, into the lower chords; the braces are set into notches in the vertical posts. Channels are cut into the inner chord planks and into the faces of the braces and vertical posts where the counterbraces cross—the posts, braces, and chords are locked together into one unit by the counterbraces.

Jan Lewandoski explains how the Paddleford truss works and how it differs from Long's truss:

"The Paddleford truss superficially resembles the Long truss. They both use a parallel chord system with vertical posts, and they both use braces and counterbraces. But the designers each had something different in mind. Long designed his truss to implement his pre-stressing idea, while Paddleford tried to make the braces work both in compression and tension by having the braces lap over all of the frame members. He tried to use wood as people later used iron rods [see Pratt truss]. It's hard to make a tension joint with wood unless you have a lot of distance to do it in. It is complicated joinery."

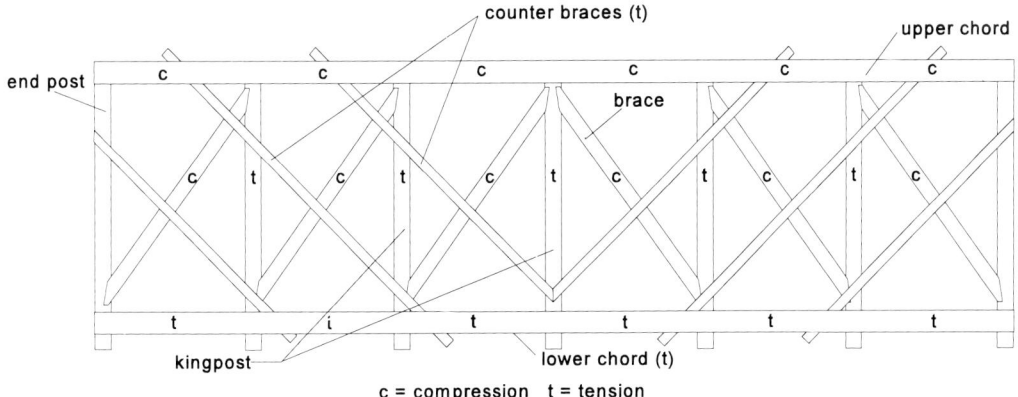

Paddleford truss

Paddleford did it with the counterbraces by running them through the series of tight-fitting channels across members that are under loading stresses. The stresses tend to re-align the truss members, causing them to lock onto the counterbrace.

The vertical posts in the 86-foot Black River Bridge in Coventry measure eight by nine inches. The braces are six-by-eight-inch timbers and the counterbraces four by six inches.

The Howe Truss

William Howe patented a wood and iron rod truss in 1840 and extended the patent with improvements in 1850. The truss was the first to be designed using mathematical stress analysis. The Howe truss was adopted by the railroad industry and became one of the most widely used trusses for railroad bridges.

The truss consists of wooden upper and lower chords, the chords linked together with sets of dual iron rods and wooden braces and counterbraces. The braces and counterbraces are butted against the chords on angle blocks. The rods are adjusted with nuts.

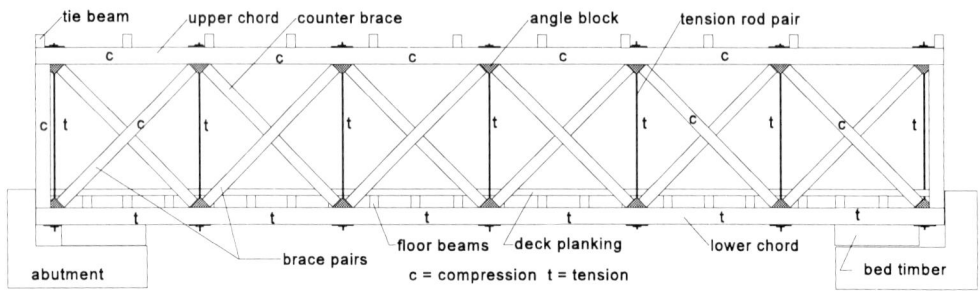

Howe Truss

Of the four Howe truss bridges in Vermont, three are highway bridges. The Gold Brook Bridge in Stowe, built in 1844, is the oldest and is a perfect miniature of the other three. The retired Rutland Railroad Bridge in Shoreham is the last of Vermont's Howe truss railroad spans. The Connecticut River bridges in Lunenburg and Lemington are constructed with identical components, both having been built by the same company. The iron rods are one-and-three-quarter-inches in diameter, the braces measure seven by nine inches, and the counterbraces are six by six inches. The angle blocks are wood.

The Haupt Truss

The Sayers Bridge in Thetford is said to be the only example of the Haupt truss in Vermont.[2] Curiously, the Sayers lacks the defining lattice structure of the braces shown in Haupt's patent drawings. Instead, it has a long and high timber arch similar in construction to Burr's arch.

The chords are constructed with two four-by-eight-inch planks sandwiching the vertical members and the braces. The vertical members and braces are four-by-eight-inch timbers pegged together between the chord planks with their long dimensions parallel to the bridge span. The arch segments are four-by-twelve-inch timbers treenailed to both sides of the

---

[2]Jan Lewandoski: "Some people think [the Sayers Bridge] is a Haupt truss, mostly because some of the diagonals cross more than one panel, but there is no evidence that the builder knew he was building a Haupt truss."

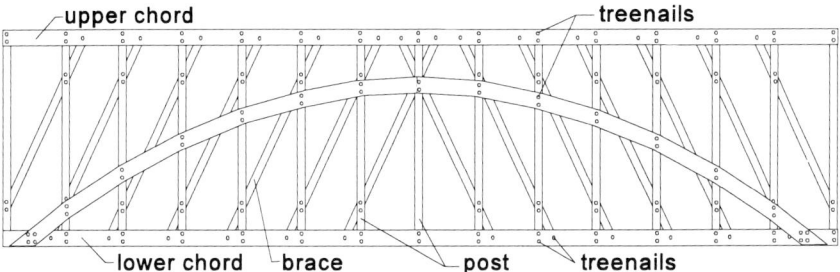

Haupt truss with arch (based on Sayers Bridge in Thetford, truss viewed fron inside of bridge)

vertical members. The open spaces between the chords and between the arch segments are filled with lengths of wood, and the whole structure is pegged together with treenails.

The Sayers Bridge spans 129 feet. In 1963 the roadway was reinforced with steel girders and a mid-stream pier.

The Pratt Truss

The Pratt truss was patented in 1844 by Thomas and Caleb Pratt. It featured panels constructed with wooden vertical posts and crossed iron-rod braces between parallel wooden chords. It was not popular in its original form, but later, when constructed entirely of iron or steel, its form became a standard for trusses.

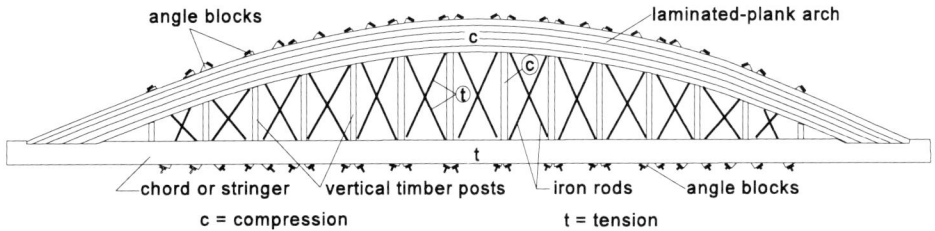

Pratt truss (based on Lincoln Bridge in Woodstock)

The Pratt truss, or one similar to it, is found in Vermont only in the 136-foot Lincoln Bridge in Woodstock, supposedly the only one of its kind implemented in wood anywhere. It is apparently called a Pratt truss because it has wooden vertical posts and crossed iron braces. This specimen, instead of having a simple wooden upper chord, is constructed with a large laminated arch.

The arch is constructed from six three-by-fourteen-inch planks. The vertical posts are three-by-fourteen-inch planks. The inch-thick iron rods are thrust through angled holes drilled through the arch—the rods cross below the arch to support the lower chord. The rod ends, upper and lower, are seated in iron angle blocks.

The Tied-Arch Truss

No one knows who designed the tied-arch truss. There are three of them in Vermont—Bests Bridge and Bowers Bridge in West Windsor and the Lake Shore Bridge in Charlotte. They are all small bridges, spanning an average of forty-one feet. The West Windsor bridges are built of a laminated arch made from five layers of two-by-ten-inch planks bent over two-by-ten-inch vertical posts. The ends of the arches are fitted into slots in the ends of the bridge chords, tying the ends of the arches together, hence the name. Three-quarter-inch rods are hung from the arch to support the bridge stringers below. Strikingly simple, the truss is a miniature model of a true arch-bridge.

Tied-arch truss

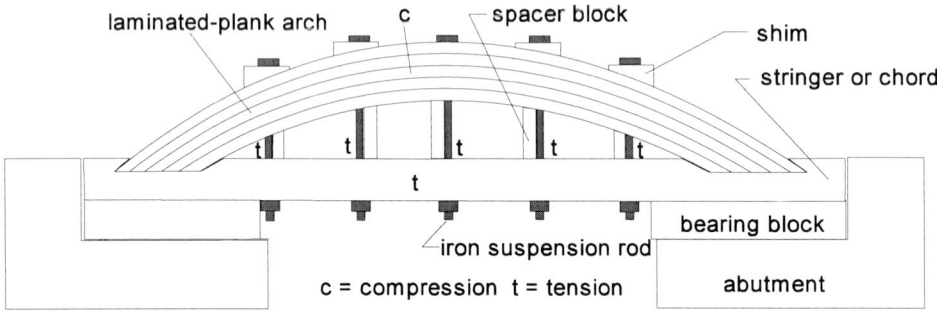

The BRIDGE BUILDERS

Appendix D

Burr, Theodore (b. 1771, Torrington, CT; d. 1822, Middletown, PA)

Theodore Burr, whose father was a millwright, apprenticed in the building trades. Soon after the turn of the century, he left Connecticut for Oxford, New York, where he built a sawmill, a gristmill, and a bridge. He followed this beginning with the building of a drawbridge in Catskill, a bridge over the Hudson River at Fort Miller, and another at Canajoharie on the Mohawk River.

In 1804 Burr built a 176-foot "arch" bridge over the Hudson River connecting Waterford and Lansingburgh, New York. He patented this design, the Burr arch, in 1817. Burr moved to Pennsylvania in 1812 and built five bridges over the Susquehanna River, all at one time. The project proved too ambitious financially, and Burr died bankrupt in 1822.

Fletcher, Bela J. (b. 1811; d. 1877, Claremont, NH)

Bela Fletcher is best known in Vermont for his work with James F. Tasker on the great Cornish-Windsor Bridge spanning the Connecticut River in 1866.

Hale, Enoch (b. 1733; d. 1813)

Colonel Enoch Hale built the first bridge in New England with a clear span longer than the length of one timber. Completed in 1785, the 365-foot bridge crossed the Connecticut River in two spans, connecting Rockingham, Vermont, with Walpole, New Hampshire.

Haupt, Herman (b. 1817, Philadelphia, PA; d. 1905, Jersey City, NJ)

Herman Haupt graduated from West Point in 1835. He resigned his commission to become district superintendent and chief engineer for the Pennsylvania Railroad. In 1839 he designed and patented the Haupt bridge truss. When the Civil War began, he was drafted to serve as superintendent of military railroads and rose to the rank of major general.

Howe, William (b. 1803, Spencer, MA; d. 1852, Springfield, MA)

William Howe invented the first truss to undergo a complete mathematical stress analysis. The Howe truss design uses timber chords and braces with iron-rod tension members. Patented in 1840 and improved in 1850, the truss became the most frequently used railroad bridge truss of its time.

Long, Stephen H. (b. 1784, Hopkinton, NH; d. 1864, Alton, IL)

An explorer, railroad engineer, and officer in the U.S. Army Engineers, Stephen Long attained the rank of brevet-colonel. In 1817 and in 1823 he led expeditions exploring the upper Mississippi River, the Rocky Mountains, Minnesota, and the northern national border. Long's Peak, near Denver, Colorado, is named for him. In 1830 Long patented the Long truss, which is based on crossed wooden braces supporting vertical wooden posts, a forerunner of the panel-type trusses used in iron bridges.

Paddleford, Peter (b. 1785, Enfield, NH; d. 1859, Littleton, NH)

Peter Paddleford was a builder of railroad bridges, mainly using the Long truss. He eventually designed his own truss based on his years of experience in building wooden bridges. The Paddleford truss design was never patented, but it was widely used and was especially popular in northeast Vermont.

Palladio, Andrea (b. 1518, Vincenza, Italy; d. 1580, Vincenza, Italy)

Andrea Palladio was an architect in the time of the Italian Renaissance. He was a proponent of classical Roman architecture, and his written work *Four Books on Architecture* (1570) influenced architects in England and the English colonies in the 1700s. Palladio's writings also included studies on timber arches and trusses.

Palmer, Timothy (b. 1751, Rowley, MA; d. 1821, Newburyport, MA)

Timothy Palmer is believed to be the first American bridge builder to advocate roofing bridges against the weather, estimating that the life of the structure would be extended by as much as forty years. A self-taught civil engineer, he built bridges in New England and Pennsylvania, including the "Permanent Bridge" across the Schuylkill River at Philadelphia. Using arch trusses, it had an overall length of 495 feet in three spans and was regarded as a masterpiece of bridge building. Built in 1804–1806, it burned in 1875.

Powers, Nicholas M. (b. 1817, Pittsford, VT; d. 1897, Clarendon, VT)

Raised on a farm near Pittsford, Vermont, Nicholas Powers apprenticed under bridge builder Abraham Owen. He built his first bridge before he was twenty-one years old and continued to build bridges in his home area until the 1850s, when he helped build the Connecticut River bridge at Bellows Falls.

In 1854 Powers built a 228-foot, single-span two-lane bridge at Blenheim, New York, using a design similar to Long's truss. In 1866 he built a railroad bridge over the Susquehanna River at Perryville, Maryland. He returned to Vermont to work as a division engineer for the Bennington & Rutland Railroad and continued to build bridges, returning to the plank-lattice truss he used when he began his career. He built his last bridge in Shrewsbury over the Cold River in 1880.

Pratt, Thomas Willis (b. 1812, Boston, MA; d. 1875, Boston, MA)

Thomas Pratt, son of Boston architect Caleb Pratt, was educated as an engineer. In 1844 he and his father patented the Pratt truss. Featuring a combination of timber posts and crossed iron diagonals, the truss was found to be difficult and expensive to construct with wood, but it later became the basis for the design of all-iron truss work. Pratt also adapted Ithiel Town's plank-lattice truss for railroad use, doubling the web and using a heavier plank. The resulting truss is called the Town-Pratt.

Tasker, James F. (b. 1826, Cornish, NH; d. 1903, Claremont, NH)

James Tasker, best known for his work in 1866 on the great Cornish-Windsor Bridge spanning the Connecticut River, was a prolific bridge builder in east-central Vermont. In his construction, he used either a plank-lattice truss or a multiple-kingpost truss of his own design. "The real genius of this self-made rural engineer," wrote Richard Sanders Allen in his *Rare Old Covered Bridges of Windsor County,* "is shown in the fact that [he] . . . could neither read nor write!"

Town, Ithiel (b. 1784, Thompson, CT; d. 1844, New Haven, CT)

Ithiel Town was an architect of public buildings and churches in New Haven, Connecticut. He designed a plank-lattice bridge truss and patented it in 1820 and again in 1835. While he was not known as a bridge builder, he sold the use of his design to builders of bridges through agents for a fee of one dollar per foot.

A COVERED BRIDGE READING LIST

Appendix E

Allen, Richard Sanders. *Covered Bridges of the Northeast*. Brattleboro, VT: Stephen Greene Press, 1957.
[This work has an extensive bibliography, a glossary, a roster of bridge builders, and a listing of all known covered bridges in New England. Allen is considered one of Vermont's leading bridge historians.]

Allen, Richard Sanders. *Rare Old Covered Bridges of Windsor County*. Brattleboro, VT: Stephen Greene Press, 1962.
[No bibliography, but includes acknowledgments.]

Branthoover, W.R. et al. *Montgomery, Vermont: The History of a Town*. Burlington, VT: Queen City Printers, 1976.
[Well-written history of the town, its people, mills, and covered bridges.]

Congdon, Herbert Wheaton. *The Covered Bridge*. Middlebury, VT: Vermont Books, 1970.
[Congdon is considered one of Vermont's leading bridge historians.]

Dummerston: An "Equivalent Lands" Town 1753-1986. Dummerston, VT: Dummerston Historical Society, 1990.
[Includes short history of the Dummerston covered bridges.]

Fairfax, Vermont: Its Creation and Development, 1776 to 1976. Bicentennial Committee, 1976.
[Well-written history of the town, its people, mills, and covered bridges.]

Fisher, Harriet et al. *Look Around Lyndon.* Lyndon, VT: Lyndon Bicentennial Booklet Committee, 1978.

[Some detail on covered bridges and people involved.]

Graton, Milton S. *The Last of the Covered Bridge Builders.* Plymouth, NH: Clifford-Nicol, Inc. 1978.

[A craftsman's view of covered bridges.]

Hagerman, Robert L. *Covered Bridges of Lamoille County.* Essex Junction, VT: Essex Publishing Co., 1972.

[Covers Lamoille River North Branch bridges not covered elsewhere.]

Harlow, Lewis A. *Covered Bridges Can Talk.* Coral Gables FL: Wake-Brook House, 1963.

[Includes seventy-five 35mm black and white photographs of Vermont's covered bridges taken in the 1940s, some of which are rare. Each bridge is described briefly, but there is no technical discussion.]

Hayes, Lyman S. *Connecticut River Valley in Southern Vermont and New Hampshire.* Rutland VT: Tuttle Company, 1929.

[A collection of short articles giving useful history and anecdotes of the area covered by the title. This book has no bibliography, attributions, or index.]

Hemenway, Abby M. *The Vermont Historical Gazetteer, Volumes I-VII.* Burlington, VT: A .M. Hemenway, 1868.

[Covers the history of Vermont towns and villages, but there is virtually no mention of covered bridges as such.]

In Sight of Ye Great River. Hartland, VT: Hartland Historical Society, 1991.

[History includes the rise of industry, plus a little text about local covered bridges.]

Johnson, Luther B. V*ermont in Flood Time-November 1927.* Randolph, VT: Roy L. Johnson Co., 1928.

[Contains anecdotes of the great flood and photographs of the aftermath.]

Morse, Victor. *Windham County's Famous Covered Bridges.* Brattleboro, VT: Stephen Greene Press, 1960.

[Introduction by Richard Sanders Allen. The text is based on research done by V. Morse in the 1930s, but there is no bibliography. Morse is regarded as one of Vermont's leading bridge historians.]

Noble, Winona S. *The History Of Cambridge, Vermont.* Cambridge, VT: Town of Cambridge, 1976.

[A history of the town mainly transcribed from town records. Covers people, places, and early bridges.]

Northfield in the Bicentennial Year 1976. Northfield, VT: Northfield Bicentennial Committee,1976.

[Examines the covered bridges and the rise of water powered mills in Northfield.]

Ritter, Michael A. *Timber Bridges, Design, Construction, Inspection, and Maintenance.* Washington DC: United State Department of Agriculture, 1990.

Shores, Venila L. et al. *Lyndon, Gem In The Green.* Lyndonville, VT: Town of Lyndon, 1986.

[Includes a short history of Lyndon's covered bridges and some of the people who built them.]

Spargo, John. *Covered Wooden Bridges of Bennington and Vicinity.* Bennington, VT: Bennington Historical Museum and Art Gallery, 1953.

[Spargo discusses the histories of the bridges of Bennington County and profiles the men who built wooden bridges. The booklet features small cuts of paintings and photographs.]

Struik, Dirk J. *Yankee Science in the Making*, New York, NY: Little, Brown and Company, 1948.

[Chapter four contains a section on the whys and wherefores of early North American economic development and the role of new science, including a discussion of wooden bridges and their builders.]

Wagemann, Clara. *Covered Bridges of New England.* Rutland, VT: Tuttle Company, 1931.

[The text is illustrated with etchings by George T. Plowman.]

Wells, Rosalie. *Covered Bridges in America.* New York, NY: William Edwin Rudge, 1931. [Contains photographs of 135 bridges in twenty-five states, but none of them are from Vermont. The rare photographs are valuable but not comprehensive, and the text is of no value to bridge researchers.]

Ziegler, Phil. *Sentinels of Time, Vermont's Covered Bridges.* Camden, ME: Down East Books, 1983.

[Features more than one hundred black & white drawings of Vermont covered bridges, with commentary.]

Atlases and Maps

The Vermont Atlas and Gazetteer. Freeport, ME: DeLorme Mapping, various editions

The Vermont Road Atlas and Guide. Burlington, VT: Northern Cartographic, Inc., various editions

Vermont City Maps. Burlington, VT: Northern Cartographic, Inc., 1993

Vermont Official State Map and Touring Guide. Vermont Agency of Transportation and Vermont Agency of Development and Community Affairs.

INDEX

abutments 52–53
Adams, Arthur 146
Allen, Denis 171
Allen, Ethan 30, 109, 188, 197
Allen, Ethan, and the Guilford Yorkers 197
Allen, Ira 197
Allen Jr., Ira 84
America-How-Are-You (TV series) 44
Amsden Village 168
arches, laminated 16, 156, 158, 190
Arlington 3, 4
Arlington Green 4
Arnold, Dr. Jonathan 101
ASA Properties Vermont, Inc. 160, 165
Ascutney Bridge, West Windsor (Smith Bridge) 166, 222
Averill, Samuel 94

Babbitt, Charles 94, 95
Bagley, Walter 123, 124
Baltimore Bridge 171–172, *172*, 222
Barber Park Bridge (Hall Bridge) 181, 224
Barrington Road Bridge (Randall Bridge) 212
Bartonsville 175
Bartonsville Bridge 175, 176, 177, *177*, 224
Batten Kill 4

Bedell Bridge 93
Bellows, Benjamin 176
Bellows Falls 175
Belvidere 43, 54, 57, 62–63
Bennington 3
Bennington Falls Bridge (Paper Mill Bridge) 200
Bests Bridge 36–37, 165, 167, *167*, 222
Big Bridge, Cambridge (Museum Bridge) 15, 48–49, 204
Big Eddy Bridge 122, *123*, 214
Big Falls, North Troy 82
Billings, Avery 24
Black Creek, Fairfield 43
Black Falls Bridge (Fuller Bridge) 71, 208
Black Falls Brook, Montgomery 71
Black River, Irasburg 83
Black River, Weathersfield 169
Black River Bridge, Irasburg 83–84, *82*, 210
Blue Bridge (Gifford Bridge) 137, 218
Boozan, Roger 49
Bowers Bridge 36, 164–165, *165*, 222
Bradley Bridge (Miller's Run Bridge) 212
Braley Bridge (Johnson Bridge) 136, 218
Brandon 20

Brattle, William 194
Brattleboro 175, 187, 194
Brewster River 51
Bridge Too Far (Avery Billings) 24
Bridge-at-the-Green 3, 4–5, 6, *4*, 200
bridge truss, definition 241–242
Brockway Mills Gorge 178
Brockway's Mills 176
Brown, Caroline, Westford Historical Society 45
Brown Bridge 28–29, *29*, 200
Browns River 43
Browns River Bridge (Westford Bridge) 208
Brownsville Bridge (Bowers Bridge) 164, 222
Bryant Bridge (Grist Mill Bridge) 52, 206
Buel, Major Elias 82
Builders
    Adams, Arthur 146
    Allen, Denis 171
    Babbitt, Charles 94, 95
    Bagley, Walter 123, 124
    Chamberlin, Harrison 179, 188, 189
    Clement, John 102, 103
    Coburn, Larned 117
    Colton, John D. 83, 84
    Emmons III, Solomon 155
    Fletcher, Bela J. 97
    French, Jason 53

265

Builders (*continued*)
　Fuller, Roscoe 53
　Goodell, Lee 102
　Granger, Sanford 176, 177, 178, 180
　Graton, Milton 37, 44, 105, 106–107, 121, 122, 144, 156, 168, 170, 171, 180
　Hale, Col. Enoch 181
　Haupt, Herman 150
　Heath, W.W. 105
　Hill, W.B. and Co., Tilton NH 34
　Holmes, George Washington 49, 53
　Ide, Paul 37, 39, 64, 122, 124, 250
　Jewett, Savanna and Sheldon 67
　Jones, J.C. 102
　Lamson, Caleb 191, 192
　Leland, Granville 171
　Leonard, Charles 65
　Lewandoski, Jan 16, 18, 37, 39, 47, 64, 74, 122, 124, 150, 155, 242, 250, 251, 253, 254
　Miller, Justin 40
　Moseley, John 132
　Mudget, Ira 140
　Newbury, Gilbert 10, 29, 51, 145, 250
　Norton, T.K. 30
　Nourse, Asa 22
　Oatman, Daniel 6
　Owen, Abraham 22, 26
　Pinney, R.W. & B.H. 157
　Pratt, Thomas & Caleb 157
　Pratt, Thomas T. 89
　Robinson, Lewis 64
　Sears, Benjamin 10
　Sears, Charles F. 10
　Sherman, Leonard 36
　Smith, John W. 115
　Stone, E.H. 107
　Stratton, Aubrey 179
　Swallows, A.W. 167
　Tasker, James H. 97, 163, 168, 170
　Tenny, Chauncey 140
　Townsend, Herman F. 119
　Tracy, Fred 65

Builders (*continued*)
　Wells, Edward 140
　Wetherby, Farewell 34, 48
　Wheeler, Eugene P. 191
　Wright, A.W. 194
Burden Iron Co. 7
Burling, Edward 45
Burr, Theodore 49, 246, 257
Burr arch truss 34–35, 246–248, *240*, *246*
Burr manner 16–17, 34

C.K. Smith Bridge (Gifford Bridge) 137, 218
Cambridge 34, 43, 48, 54–55
Cambridge Center 48
Cambridge Junction 48, 53
Cambridge Junction Bridge 53–55, *54*, 206
Canyon Bridge, Cambridge (Grist Mill Bridge) 206
Captain Joe and Molly 111
Cedar Swamp Bridge (Salisbury Station Bridge) 202
Cemetery Bridge (Coburn Bridge) 214
Center Bridge (Sanborn Bridge) 212
Chaffee Bridge (Red Bridge) 114, 216
Chamberlin, Ephraim 105
Chamberlin, Harrison 179, 188, 189
Chamberlin, Myron 105
Chamberlin Bridge 101, 103, 104–105, *104*, 212
Chamberlin Mill Bridge (Chamberlin Bridge) 104, 212
Chapel-on-the-Green 4
Charlotte 33
Chelsea 135, 145–146
Chiselville Bridge 3, 5–6, *5*, 200
Chittenden, Gov. Thomas B. 119
Church Street Bridge 58–59, *59*, 206
Cilley Bridge 135, 141–142, *141*, 216
Clarendon 29–30
Clement, John 102, 103
Coburn, Larned 117
Coburn, Harry L. and F.W. 118
Coburn Bridge 116, 117–118,

Coburn Bridge (*continued*)
　118, 214
Codding Hollow Bridge (Jaynes Bridge) 61, 206
Coit's Gore 58
Cold River 28
Colton, John D. 83, 84
Columbia Bridge 94–95, *95*, 212
Comstock, John 72
Comstock Bridge 70, 72, *73*, 208
Connecticut River 93, 96
Conner, Bernard 56
Cooley, Gideon and Benjamin 25, 26
Cooley Bridge 22, 24, 25–26, *25*, 200
Cornish-Windsor Bridge 97–99, *97*, 212
Cornwall 18
Coventry 82–83
Coventry Bridge (Black River Bridge) 82, 210
Covered Bridges and the Three R's 62
Cox Brook 127, 128
Creamery Bridge, Brattleboro 187, 194, *195*, 226
Creamery Bridge, Montgomery 70, 73–74, *74*, 208
cribs, timber 16
cross-x bridge 46
Crystal Palaces 72
Crystal Springs Bridge (Creamery Bridge, Montgomery) 208
Cumberland County, New York 195
Cummings, H.P., Construction Company 159, 166

Danville 101, 109–110
D'Anville, Admiral Le Duc 109
Dean Bridge 22
Deerfield, Connecticut, raid 176
Depot Bridge 22, 24, *25*, 202
Dog River 127, 128
Doolittle, Captain Ephraim 25
double-barrel bridge 15
Douglas, Stephen A. 20
Downers Bridge 21, 167, 169–170, *170*, 222
Dummer, William 194
Dummerston 187
Dunmore, Governor John 30
Durham Township, Clarendon 29

East Berkshire 76
East Branch of the Passumpsic River 108
East Creek 27
East Fairfield 55
East Fairfield Bridge 55–56, *56*, 206
East Montpelier 113, 117
East Randolph 135
Eden 43, 86
Egypt Road 57
Emily's Bridge (Gold Brook Bridge) 115, 214
Emmons, Edwin 155
Emmons III, Solomon 155
Enos, Roger 75
Enosburg 75–76
Eureka School House 171
Everts, John 13
ex-bridge 46

Fairfax 43, 45–48
Fairfield 43
Ferris, Benjamin 33
First Bridge Across the Connecticut 181
Fisher, Christopher 90
Fisher Bridge 89–90, *90*, 210
Flanders, Senator Ralf E. 171
Fletcher, Bela J. 97, 257
Flint Bridge 135, 144–145, *145*, 218
floods
    of 1806 120
    of 1814 46
    of 1824 120
    of 1826 192
    of 1830 120, 122
    of 1832 46
    of 1839 192
    of 1869 5, 177, 182, 188, 192
    of 1927 23, 46, 47, 103, 105, 106, 118
Florence 22
Florence Station Bridge 24
flying buttresses 80, 180–181, 182, 189
Foster, Nathaniel 57
French, Jason 53
Fuller, Roscoe 53
Fuller Bridge 70, 71–72, *71*, 208
Furnace Brook 25

Garfield Bridge, Hyde Park 159, 165
Gates Farm 48–49
Gates Farm Bridge 50–51, *50*, 206
Gifford Bridge 135, 137–138, *138*, 218
Gihon River 86, 88
Girardin, Susan 62
glulaminated beams 97, 183
Gold Brook 115, 117
Gold Brook Bridge 115–116, *116*, 214
Gold in the Brook 117
Goodell, Lee 102
Goodnough Bridge (Gorham Bridge) 202
Gorham Bridge 22, 26–27, *27*, 202
Grafton 175, 182
Grand Canyon Bridge (Grist Mill Bridge) 52, 206
Granger, Sanford 176, 177, 178, 180
Graton, Milton 37, 44, 105, 106–107, 121, 122, 144, 156, 168, 170, 171, 180
Great Cedar Swamp 17
Great Eddy Bridge (Big Eddy Bridge) 214
Great Falls 46
Greenbanks Hollow Bridge 101, 110–111, *110*, 212
Green Mountain Boys 7, 29–30
Green River Bridge 187, 195–196, *196*, 226
Greensboro 43
Grist Mill Bridge 51–53, *52*, 206
Guildhall 94
Guilford 187, 195
Guy Bridge (Moxley Bridge) 146, 218

Hale, Colonel Enoch 181, 257
Hale's Bridge 181
Hall Bridge 180–181, *180*, 224
Halpin Bridge 13, 14–15, *14*, 202
Hammond Bridge 22, 23–24, 23, 202
Harnois Bridge (Longley) 210
Hartland 163
Haupt, Herman 150, 258

Haupt truss 150
    description 254–255, *255*
Hayward and Noble Bridge (Mill Bridge) 143, 218
Hazeltine, Col. John 188
Head Bridge (Longley Bridge) 74, 210
Heath, W.W. 105
Hectorville Bridge 68, *69*, 208
Henry, Berntine T. 8
Henry, William 7, 8
Henry Bridge 3, 6–8, *8*, 200
Herrins, John & Robert, woolen mill 58
highest covered bridge 14
Hill, W.B. and Co., Tilton NH 34
Historical Society, Weathersfield 168
Historical Society, Westford 44
Holmes, George Washington 49, 53
Holmes Creek 36
Holmes Creek Bridge (Lake Shore Bridge) 36, 204
Hopkins Bridge 75–76, *76*, 208
Howe, William 253, 258
Howe Bridge 135, 140–141, *140*, 218
Howe truss 94, 115–116
    description 253–254, *254*
Howrigan, Francis & Michael 56
Hutchins, Joseph 69
Hutchins Bridge 69–70, *70*, 210
Hyde Bridge 135, 138–139, *139*, 218

Ide, Paul 37, 39, 64, 122, 124, 250
Irasburg 83, 84
ISTEA 44

Jaynes Bridge 58, 61–62, *61*, 206
Jeffersonville 48, 49, 52
Jewett, Savanna and Sheldon 67
Joe's Brook 110, 111
Joe's Pond 111
Johnson 86, 88
Johnson, Samuel William 86
Johnson Bridge 135, 136–137, *137*, 218
Johnson Woolen Mills, The 86
Jones, J.C. 102

INDEX

267

Junction Bridge, Belvidere
(Lumber Mill Bridge) 206
Junction Bridge, Cambridge 53
Justin Morgan Memorial 145

Kelly, John 62
Kelly River 62
Kennedy, Tom, Westford
Historical. Soc. 44
Kidder Hill Bridge 175, 182–
183, *183*, 224
king-rods 180, 182, 184, 188
Kingpost truss, description 243–
244, *243*
Kingsbury and Stone 47
Kingsbury Bridge (Hyde Bridge)
138, 218
Kingsley, John H., Mr. & Mrs. 30
Kingsley Bridge 30, *31*, 202
Kipling, Rudyard 192
Kissing Bridge (Jaynes Bridge)
61, 206

Lake Shore Bridge 33, 36–37,
36, 165, 204
laminated wood beam 97, 183
Lamoille River 43, 46
Lamoille River, North Branch
43, 57, 62
Lamoille River Bridge 46
Lamson, Caleb 191, 192
Larkin Bridge 135, 143–144,
144, 218
Leland, Granville 171
Lemington 93, 94
Lemon Fair River 19, 20
Lemon Fair River, name origin 20
Leonard, Charles 65
Lewandoski, Jan 16, 18, 37, 39,
47, 64, 74, 122, 124,
150, 155, 242, 250, 251,
253, 254
Lewis Creek 37
Lewis Creek valley 38
Lincoln, Abraham 20
Lincoln Bridge 89, 154, 157–
158, *158*, 220
Lincoln Gap Bridge 123–124,
124, 214
Little Bridge (Gates Farm
Bridge) 48–51, 206
Locust Grove Bridge (Silk Road
Bridge) 11, 200

Long, Stephen H. 251, 258
Long Truss, description 251–
252, *252*
longest covered bridge in
Vermont
in two spans 193
under one roof 193
single span, plank-lattice 177
longest wooden bridge in U.S. 97
Longley Bridge 70, 74–75, *75*, 210
Lords Creek 84
Lords Creek Bridge 83, 84–85,
85, 210
Lower Bridge, Belvidere
(Lumber Mill Bridge) 206
Lower Bridge, Brandon (Sanderson Bridge) 22, 204
Lower Bridge, Charlotte
(Quinlan Bridge) 204
Lower Bridge, Fairfax (Maple
Street Bridge) 206
Lower Bridge, Irasburg (Black
River Bridge) 83, 210
Lower Bridge, Tunbridge (Cilley
Bridge) 141, 216
Lower Bridge, Waterville
(Montgomery Bridge) 60,
208
Lower Cox Brook Bridge
(Newell Bridge) 129, 216
lowest covered bridge 36
Lulls Brook 163
Lumber Mill Bridge 58, 63–64,
63, 206
Lunenburg 93, 95
Lunenburg-to-Lancaster Bridge
(Mount Orne Bridge) 95,
212
Lydius, Col. John Henry 29
Lyndon, Josias 101
Lyndon 101–102
Lyndon Center 106
Lyndon Corner 105
Lyndon Cutter 105
Lyndonville 102

MacMillan Bridge 175, 184,
184, 224
Mad River 113, 120, 122, 123
Maple Street Bridge 46–48, *47*,
206
Marsh, Capt. Isaac 119
Marshfield 113, 119

Martin, William 119
Martin Bridge 118, 119, *120*, 214
Martin's Mill Bridge 163–164,
164, 222
Mead Bridge 22
Middle Bridge, Northfield 130
Middle Bridge, Woodstock 154,
156–157, *157*, 220
Middlebury 13
Millbank, R.V. 34
Mill Bridge, Belvidere (Lumber
Mill Bridge) 206
Mill Bridge, Tunbridge 135,
142–143, *142*, 218
Mill Brook, Fairfax 46
Mill Brook, West Windsor 164
Miller, Anson 105
Miller, Justin 40
Miller's Run Bridge 101, 107–
108, 111, *108*, 212
Mill River 30
Mill River Bridge (Kingsley
Bridge) 202
Missisquoi River 43, 79
Montgomery, Dallas 60
Montgomery, Gen Richard 67
Montgomery, Town of 67
Montgomery Bridge 58, 59–60,
60, 208
Montgomery Center 68
Montgomery Village 68, 70
Montpelier 136
Morgan Bridge 63, 64–65, *64*,
208
Morgan Horse, The 141
Morristown 113
Morrisville 113
Moseley, John 132
Moseley Bridge 129, 132, *133*,
216
Mount Orne Bridge 95–96, *96*,
212
Moxley Bridge 135, 146–147,
146, 218
Muddy Branch of the New
Haven River 14
Mudget, Ira 140
Mudget Bridge (Scribner Bridge)
86, 210
multiple-kingpost half-high
truss 136
multiple-kingpost plus Burr
arch truss 34

multiple-kingpost truss, description 245, *245*
Museum Bridge 33, 34–35, *35*, 204

narrowest covered bridge 181
National Geographic Society 44
Nelson and Hall Co. 68
neo-Grecian 21
Neshobe 20
Newbury 93
Newbury, Gilbert 10, 29, 51, 145, 247, 250
Newell Bridge 128, 129–130, *130*, 216
Newfane 187, 190
Noble, Obadiah 139
North Bennington 3, 6, 8
North Branch of the Lamoille River 43, 57, 59, 60, 61, 62, 63, 65
North Branch of the Passumpsic River 104
Northfield 127–129
Northfield Falls 128
Northfield Falls Bridge (Station Bridge) 129, 216
North Ferrisburgh 38
North Hartland 149, 152
North Troy 79
Norton, T.K. 30
Nourse, Asa 22

Oatman, Daniel 6
oldest covered bridge 13
Old Sturbridge Village, Massachusetts 193
Ompompanoosuc River 149
open bridges 136
Orne Bridge (Black River Bridge) 210
Orton Farm Bridge (Martin Bridge) 119, 214
Orton's Vermont Country Store 179
Osgood Bridge (Hall Bridge) 181, 224
Ottauquechee River 152
Ottauquechee Woolen Mill 152
Otter Creek 13, 18, 20
Owen, Abraham 22, 26

Paddleford, Peter 252–253, 258

Paddleford truss 83, 84, 85, 101, *106*
 description 252–253, *253*
Paine, Elijah 127
Palladio, Andrea 258
Palmer, Timothy 259
Paper Mill Bridge 3, 6, 9–10, *9*, 200
Passumpsic River 102
Perkinsville 170
Pine Brook Bridge 119, 120–121, *121*, 214
Pinney, R.W. & B.H. 157
Pitt, William 22
Pittsford 22–23, 26
Plainfield Village 119
plank-lattice/kingpost hybrid 68
plank-lattice truss, description 248–251, *249*
plank-lattice truss tested 4
Poland, Judge Luke P. 54–55, 58
Poland, Luther 58
Poland Bridge (Cambridge Junction Bridge) 53, 206
Pomfret Bridge (Smith Bridge) 160, 220
Potter Bridge (Jaynes Bridge) 60, 208
Power House Bridge 87, 88–89, *88*, 210
Powers, Nicholas M. 22–23, 25, 26, 27, 28, 29, 259
Pratt Construction Company 89
Pratt, Thomas and Caleb 157, 255
Pratt, Thomas W. 89, 259
Pratt truss, description 255–256, *255*
Proctor 26
Pulp Mill Bridge 13, 15–17, 35, 122, *15*, 202
Pumpkin Harbor 49

queenpost truss, description 244–245, *244*
Quinlan Bridge 37–38, 39, *38*, 204

Railroads
 Green Mountain 177
 New England Central 128
 St. Johnsbury & Lamoille County 53, 89

Randall Bridge 101, 108–109, *109*, 212
Randolph 135
 and the State Capitol 135
Randolph, Peyton 135
Red Bridge 113, 114–115, *114*, 216
River Road Bridge (School House Bridge) 210
Roaring Branch of the Batten Kill 5
Robinson, Lewis 64
Robinson Bridge (Silk Road Bridge) 11, 200
Rockingham 175, 176
Rockwell, Norman 4
Root, Abner 139
Rutland 27
Rutland Railroad Bridge 19–20, 51, *19*, 202

Salisbury 13, 18
Salisbury Station Bridge 17–18, *17*, 202
Salmond Bridge 168–169, *169*, 222
Samuel Head Bridge (Longley Bridge) 74, 210
Sanborn, Benjamin 106
Sanborn, Isaac W. 102
Sanborn Bridge 101, 105–107, *106*, 212
Sanderson Bridge 20–21, *21*, 169, 204
Sawmill Bridge (Chamberlin Bridge) 104, 212
Saxton's River 175
Sayers Bridge 150–151, *150*, 220, 254–255
Scenic Roads Resource 17
School House Bridge, Lyndon 101, 102–103, 124, *103*, 214
School House Bridge, Troy 80–83, *81*, 210
School Street Bridge (Power House Bridge) 88, 210
Scott Bridge, Cambridge (Grist Mill Bridge) 52, 206
Scott Bridge, Townshend 187, 188–190, *189*, 226
Scribner Bridge 85, 86–87, *87*, 210

Sears, Benjamin 10
Sears, Charles F. 10
second-highest covered bridge 6
second-longest single-span covered bridge 54
Seguin Bridge 37, 38–39, *39*, 204
Seymour River 48, 49, 50
Shelburne 34
Shelburne Museum 34, 49
Sherman, Leonard 36
Sherman Bridge (Quinlan Bridge) 204
Sherman Brook 168
Sherman's sawmill 38
Shirley, Governor William 30
Shoreham 19
Shoreham Bridge (Rutland Railroad Bridge) 19, 202
Shrewsbury 29
Silk Road Bridge 3, 10–11, *10*, 200
sistering 47–48
skewed bridge 131, 144, 146
Slaughterhouse Bridge 129, 131–132, *132*, 216
Slayton, Capt. A.H. 117
Smith, Benjamin, of Troy, N.Y. 80
Smith, J E., bobbin factory 71
Smith, John W. 115
Smith Bridge, Brownsville 165–166, *166*, 222
Smith Bridge, Pomfret 154, 158–160, *159*, 220
snowing the bridge 136
Socialborough grant, Clarendon 30
South Branch of the Saxton's River 183
South Branch of the Trout River 68
South Wheelock Branch 103
Spade Farm Bridge 33, 40, *40*, 204
Springfield 163
Station Bridge, Northfield 128, 129, *128*, 216
Station Bridge (Cambridge Junction Bridge) 53
Station Bridge (Salisbury Station Bridge) 13, 17, 202
Sterling Brook 114
Sterling Brook Bridge (Red Bridge) 114, 216

Stone, E.H. 107
Stoney Brook 132
Stoney Brook Bridge (Moseley Bridge) 216
Story, Ann 18
Stoughton Bridge 168, 170–171, *171*, 224
Stoughton Pond 168
Stowe 113, 115
Stowe Hollow Bridge (Gold Brook Bridge) 214
Stratton, Aubrey 179
Sunderland 3, 5
Swallows, A.W. 167
Swallows Bridge (Best Bridge) 167, 222

Taft, Stephen 154
Taft Bridge, Dummerston 193
Taftsville 154
Taftsville Bridge 154, 155–156, *155*, 220
Tasker, Henry 170
Tasker, James H. 97, 163, 168, 170, 260
Taylor, Thomas 89
Tenny, Chauncey 140
Thetford 149
Thetford Center Bridge (Sayers Bridge) 220
tied-arch truss 36, 164–165
  description 256, *256*
timber cribs 16
timber-lattice truss 97
Titcomb, Andrew 170
Titcomb Bridge (Stoughton Bridge) 170, 224
Tours
  1– Bennington County Bridges 3, 200
  2– The Otter Creek Basin 13, 200–204
  3– The Wooden Bridges of Charlotte 33, 204
  4– The Lamoille River and the North Branch 43, 206–208
  5 – The Town of Montgomery 67, 208–210
  6 – Route 100 in Northern Vermont 79, 210
  7 – Crossing the Connecticut 93, 212

Tours (*continued*)
  8 – The Lyndon Bridges 101, 212–214
  9 – Route 100 in Central Vermont 113, 214–216
  10 – Northfield 127, 216
  11 – The Northern Tributaries of the White River 135, 216–218
  12 – Woodstock 149, 220
  13 – The Windsor Area 163, 222–224
  14 – Rockingham to Grafton 175, 224
  15 – The Deep South 198, 226
Town, Ithiel 248, 260
Town-Pratt truss 90
Townsend, Herman F. 119
Townshend 187, 188
Townshend Depot Bridge 179
Tracy, Fred 65
Trout River 68
Troy 79–80
Trusses, bridge
  definition and description 241–243
  Haupt 254–255, *255*
  Howe 253–254, *254*
  kingpost 243–244, *243*
  Long 251–252, *252*
  multiple-kingpost 245, *245*
  Paddleford 252–253, *253*
  plank-lattice 248–251, *249*
  Pratt 255–256, *255*
  queenpost 244–245, *244*
  tied-arch 256, *256*
Tub Factory, The, Montgomery 69
Tucker Toll Bridge 181
Tunbridge 135, 139–140
Tunbridge Worlds Fair 143
Turkey Bridge, The 154
Twigg-Smith, Jr., Thurston 160, 165
Twin Bridge 27–28, *28*, 204

U.S. Army Corps of Engineers
  Springfield-Weathersfield flood control 168
  Townshend flood control 179
Union Street Bridge (Middle Bridge) 156, 220
Union Village Bridge 151–152, *152*, 220

unique trusses 114, 156
Upper Blaisdel Bridge (Johnson Bridge) 136, 218
Upper Bridge, Belvidere (Morgan Bridge) 208
Upper Bridge, Brandon (Dean Bridge) 22
Upper Bridge, Charlotte (Seguin Bridge) 39, 204
Upper Bridge, Northfield (Upper Cox Brook Bridge) 128, 216
Upper Bridge, Troy (School House Bridge) 210
Upper Bridge, Waterville (Jaynes Bridge) 61, 206
Upper Cox Brook Bridge 130–131, *131*, 216
Upper Falls Bridge (Downers Bridge) 169, 222

Vermont Agency of Transportation (VAOT) covered bridge report
    Bests Bridge 167
    Black River Bridge 84
    Burr Truss as "multiple kingpost plus Burr" 34
    Cambridge Junction Bridge 53–55
    Chiselville Bridge 6
    Creamery Bridge 194
    Creamery Bridge, Montgomery 73
    East Fairfield Bridge 56
    Flint Bridge 145
    Gates Farm Bridge 50–51
    Gifford Bridge 138
    Greenbanks Hollow Bridge 111
    Grist Mill Bridge 53
    Halpin Bridge 14
    Henry Bridge 7, 8
    Hopkins Bridge 76
    Howe Bridge 140
    Hutchins Bridge 69
    Kidder Hill Bridge 182–183
    Lake Shore Bridge 36
    Mill Bridge, Tunbridge 143
    Morgan Bridge 65
    Paper Mill Bridge 9

Vermont Agency of Transportation (VAOT) covered bridge report *(continued)*
    Pulp Mill Bridge 16
    School House Bridge, Troy 81
    Scribner Bridge 87
    Seguin Bridge 39
    Union Village Bridge 152
    Westford (ISTEA) 44
    Williamsville Bridge 191
Vermont Market Road 43
Victorian Village Bridge 175, 176, 179–180, *179*, 224
Village Bridge, Waitsfield (Big Eddy Bridge) 214
Village Bridge, Waterville (Church Street Bridge) 58, 206

Wait, General Benjamin 119–120
Waitsfield 113, 119–120, 122
Walloomsac River 6, 11
Waltham Turnpike Company 13
Warren 113, 123
Warren Bridge (Lincoln Gap Bridge) 214
Waterville 43, 54–55, 57–58
Weathersfield 163, 167–168
Webb, Mrs. J. Watson 34
Wells, Edward 140
Wentworth, Governor Benning
    Bennington Grant 3
    Brattleboro, Dummerston, and Putney charters 194
    Charlotte grant 33
    Clarendon grant 30
    Fairfax grant 45
    Fairfield grant 55
    Fane charter 190
    Hartland grant 152
    Lemington grant 94
    Neshobe grant 20
    River of Pines 176
    Tunbridge grant 139
    Weathersfield grant 167
    Windsor grant 96
West Branch of the Passumpsic River 106
West Dummerston 187

West Dummerston Bridge 187, 188, 192–193, *193*, 226
Westford 43–45
Westford Bridge 43–45, *44*, 208
Westford Historical Society 44
West Hill 67, 73–74
West Hill Bridge (Creamery Bridge, Montgomery) 208
West Hill Brook 73
West River 175, 187
West Windsor 165
Wetherby, Farewell 34, 48
Weybridge 17
Wheeler, Eugene P. 191
Whetstone Brook 194
Whitcomb, Harold 105
Whitcomb Bridge (Chamberlin Bridge) 104, 212
Whitcomb's Mill 105
White River, First and Second Branches 135
widest covered bridge 46
Wilder Bridge (Pine Brook Bridge) 120, 214
Willard, P.K. 152
Willard Bridge 152–153, *153*, 220
Willard, Oliver 152, 154
Williams, Rev. John 176
Williams River 175, 176
Williams River Bridge (Bartonsville Bridge) 224
Williamsville 190
Williamsville Bridge 187, 190–191, *191*, 226
Windsor 93, 96–99
Windsor, birthplace of Vermont 96
Winooski River 113, 117, 118
Wolcott 89
Wolcott, Major General Oliver 89
wooden kingposts 120, 132
Woodstock 149, 154
Worrall Bridge 175, 176, 178–179, *178*, 224
Wright, A.W. 194
Wright Construction Company 158
Wrong-way Bridge 49

Young, Kerry 62